Series in BioEngineering

The Series in Bioengineering serves as an information source for a professional audience in science and technology as well as for advanced students. It covers all applications of the physical sciences and technology to medicine and the life sciences. Its scope ranges from bioengineering, biomedical and clinical engineering to biophysics, biomechanics, biomaterials, and bioinformatics.

Indexed by WTI Frankfurt eG, zbMATH.

Seyed Morteza Naghib · Yasaman Rezaeian

Frontiers of Nanobiotechnology

Microfluidic Devices and BioMEMS for the Future

Springer

Seyed Morteza Naghib
Department of Nanotechnology
Iran University of Science and Technology
Tehran, Iran

Yasaman Rezaeian
Department of Nanotechnology
Iran University of Science and Technology
Tehran, Iran

ISSN 2196-8861 ISSN 2196-887X (electronic)
Series in BioEngineering
ISBN 978-3-031-88418-4 ISBN 978-3-031-88419-1 (eBook)
https://doi.org/10.1007/978-3-031-88419-1

This Springer imprint is published by the registered company Springer Nature Switzerland AG
The registered company address is: Gewerbestrasse 11, 6330 Cham, Switzerland

Contents

Chapter 1
Manufacturing Methods for Microfluidic Devices Using Microscale Physics

Abstract Microfluidics involves the control and manipulation of fluids on a small scale, leading to unique phenomena and the development of functional components and machines. This chapter focuses on the fundamental physics of microfluidics and the standard methods used to fabricate microfluidic devices. It explores the distinctive behavior of fluids at the microscale, including low Reynolds numbers, dominant diffusion effects, and amplified surface tension forces. Understanding these phenomena is essential for designing effective microfluidic devices. The chapter also discusses various fabrication methods such as micromachining, soft lithography, molding, 3D printing, and paper-based techniques, highlighting their advantages and disadvantages. These fabrication methods play a crucial role in realizing the potential of microfluidic devices in the life sciences. By combining the understanding of microscale physics with appropriate fabrication techniques, researchers can create innovative microfluidic devices for applications in diagnostics, drug delivery, tissue engineering, and molecular analysis. This chapter serves as a comprehensive introduction to the physics of microfluidics and the fabrication methods involved, laying the groundwork for further advancements in this exciting field.

Keywords Manufacturing methods · Microfluidics · Devices · Microscale · Physics

1.1 Introduction

Microfluidics is the field of knowledge and technology focused on structures with interconnected channels on a scale ranging from hundreds to tens of micrometers [1–3]. These systems enable precise control and manipulation of small volumes of fluids, typically in the microliter to milliliter range, within designed configurations [4–7]. The emergence of microfluidics can be traced back to the need for planning and miniaturizing bio(chemical) research [8–11]. The concept of "Lab-on-a-chip"

S. M. Naghib and Y. Rezaian, *Frontiers of Nanobiotechnology*, Series in BioEngineering, https://doi.org/10.1007/978-3-031-88419-1_1

or "total analysis systems" (μTAS) at the micrometer scale gradually gained recognition [12]. At the microscale, fluids exhibit different behaviors compared to macroscopic fluids due to the increased surface area-to-volume ratio [13]. This leads to three significant phenomena: the dominance of viscous forces over inertial forces, highly efficient mass–heat transfer, and notable surface effects [4]. Additionally, the high connectivity of microfluidic channels facilitates the symbiosis and interplay of various fluid phases, forming the foundation for manipulating and controlling fluids at a small scale [14]. As a result, microfluidics has illuminated research disciplines in biology, chemistry, medicine, physics, and engineering [15].

Microfluidic devices provide several benefits over their macroscale counterparts, such as low sample and reagent consumption, quick reaction times, high surface-to-volume ratios, and cost-effectiveness [16]. The reduction in sample/reagent consumption is particularly important when working with expensive materials [17]. The ability to create patterns and structures on small length scales, down to the micron level, has spurred extensive research and the development of tools for fluid and surface manipulation and transport [18]. Engineering schemas are employed in microfluidics to achieve precise control and design at length scales approaching the molecular level [19]. Consequently, microfluidics has become an integral part of traditional fluid dynamics research [20]. Modern advancements in microfluidic systems have found numerous applications in fluid transportation, including the life sciences industry for biomedicine and pharmaceuticals. These applications encompass diagnostic devices, drug innovation, delivery, and detection [21]. Microfluidics is also utilized in industrial settings for combinatorial synthesis, enabling quick chemical analyses and high-throughput sieving [10]. Furthermore, micro-devices that can be swallowed or implanted have facilitated non-invasive sample analysis and surgical procedures in various medical fields. The research field of "nanofluidics" represents a challenge that involves contributions from continuum, statistical, and molecular mechanics [20].

Approximately 20 years ago, research on microfluidic systems utilizing micromechanics technology commenced. Early examples included the use of a gas chromatograph and IBM inkjet printer nozzles at Stanford University. While the development of these devices was initially limited, there has been a significant surge in research activities in recent years, making microfluidics a prominent and actively researched area. Modern microfluidic devices consist of various components, such as gas pressure control vents, flow sensors, sophisticated microfluidic control systems for chemical analysis, separation capillaries, pumps, and chemical detectors. These components can be mounted on a single surface or assembled into modules [22]. Microdevices and nanodevices are beneficial because they permit manipulation with rapid reaction times; they can control and sense flows, manage small fluid volumes, and template substrates on small length scales. Most importantly, they can selectively examine them at the cell scale. Typically, silicon micro-machining methods are used to build micron-scale mechanical and electrical devices from glass and silicon. However, more recently, elastomeric substances have replaced the previous ones [23]. The recent systems render nanometer-scale properties and the probability of transforming shapes [24]. Both production methods can play a key role in future applications [20].

Microfluidic systems have achieved notable advancements in specific applications in recent years, particularly in lab-on-a-chip tests and inkjet printers. Further innovations and advancements in this field have the potential to make microfluidic technology pervasive across various industries [6]. These industries include agriculture, life sciences, public health, biotechnology, defense, and pharmaceuticals, each with its unique requirements [25]. Many sectors are looking to the next generation of microfluidic devices to hit the market because of its focus on efficiency and adaptability. Professor George Whitesides of Harvard University's Departments of Chemical Biology and Chemistry notes that microfluidics seeks to construct more complicated systems of channels with advanced fluid handling capabilities, in contrast to microelectronics, which concentrates on shrinking the size of transistors. Although microfluidic devices require similar components such as separators, filters, mixers, valves, and pumps, the small size of microchannels alters fluid flow behavior [20].

Microfluidic devices have the potential to automate a wide range of biological and chemical operations, enabling more efficient diagnostic and therapeutic procedures with improved reproducibility [26–28]. Polymer-based microfluidic devices offer specific advantages, including biocompatibility and cost-effectiveness [5]. Moreover, they can be manufactured from a variety of materials using different methods [29]. Continuing our exploration, we will delve into the physical attributes of microfluidics and the manufacturing methods employed in microfluidic device production.

1.2 Exploring the Physics of Microfluidics

The study of microfluidic physics is highly significant. In this section, we will describe the standard methods used to create microfluidic systems and devices based on microfluidic physics. The focus of this study is on micro-scale phenomena and the utilization of micro-scale physics to construct systems and devices that offer valuable functions in the field of life sciences. To comprehend and utilize microfluidics effectively, it is crucial to first identify the dominant physical events that occur on a micro-scale. The components of microfluidic devices include pumps, mixers, and valves, which intelligently control fluid flow using the principles of microscale physics. However, the published microfluidic studies do not provide a comprehensive overview of small-scale physics or how specific devices can be realized. Therefore, our aim is to present a more detailed understanding of microfluid physics. Microfluidics involves the manipulation and analysis of liquids in micrometer-scale structures. While microfluidic systems initially utilized technology developed for microchip fabrication, it has now evolved into a distinct discipline.

The forces operating at a microscopic scale are different from those at a macroscopic scale [30]. Scaling down existing technologies and expecting them to work properly at the microscale frequently yields undesirable consequences [31]. As a result, novel systems that make use of the specific forces present at the microscale are required [32]. Since the mid-1990s, there has been significant progress in understanding the physics of microfluidic flows, enabling their application across various domains. Today, there are numerous comprehensive books and reviews available, focusing on the physical characteristics of fluids. These resources describe scaling rules that govern the balance of forces at the microscale, along with discussions on the characteristics of low Reynolds number flows (where viscosity dominates over inertia), diffusion, surface-to-volume ratio, and surface tension. Understanding these concepts is vital for fluid performance in complex geometries within microfluidic devices, designing efficient mixing devices, operating with microparticles, and achieving effective separation [33]. Microfluidic physics is a rich field that draws from electrostatics, classical chemistry and physics, fluid mechanics, statistical mechanics, thermodynamics, elasticity, and polymer physics. Importantly, the small dimensions at the microscale introduce different physical phenomena that are less familiar at the macroscale.

While the design and construction of microfluidic tools and devices are of primary interest to some researchers, scientific exploration and understanding of the underlying physics are the focus for others. One significant difference between microfluidics and microelectronics is that as the dimensions scale down, the fundamental physics changes more rapidly in microfluidics [34]. Despite decades of development in semiconductor electronics that have led to increased densities and smaller transistors, scaling alone has not brought about significant qualitative changes in physical phenomena, such as the transition from classical to quantum regimes. Additionally, mesoscopic technologies like single-electron transistors remain relatively unknown objects in laboratory research [35]. In contrast, fluid systems can rapidly reach length scales where the fundamental physics of the fluid undergoes substantial transformations [36]. An important example is the dominance of viscous dissipation and negligible inertia in mass transport within microfluidic systems. The nonlinearity of inertia, responsible for turbulence and instabilities in larger scales, is of minimal interest in microfluidic currents [10]. Therefore, this study focuses on the physics of microfluidic devices and the range of physical phenomena and how to exploit them. The current research in this field emphasizes its specialized applications and should complement other physical studies conducted by scientists. We primarily consider fluid physics occurring in channels with widths on the order of 100 μm (1 μm = 100 nm is used as a reference). The physics studied here is relevant to biological systems, porous media, colloids, and complex liquids. The introductory fluid physics captures the relative importance of different phenomena through a series of dimensionless numbers. We intend to examine the parameter space created by these dimensionless numbers in microfluidic physics. In some cases, these numbers are scaled by a specific system length, such as channel size, thereby representing a crucial change in physics as the volume of fluid is reduced to the nanoliter scale. Towards the end of the study, we will identify additional phenomena that characterize microfluidic devices [37].

Over the last few decades, microfluidics has witnessed significant growth, as researchers have made remarkable progress in various areas such as rapid prototyping, exploration of different channel structures, mixing of liquids and particles, and integration of optical and electrical systems to address challenges in chemical and biological analysis, healthcare, material synthesis, and other emerging fields that can benefit from the unique scale, physics of microfluidic systems, or automation. One emerging field that takes advantage of the distinctive physical phenomena found in micro-scale patterned channels is inertial microfluidics, which utilizes the unconventional utilization of fluid inertia in microfluidic operating systems.

The shape of the channel plays a crucial role in determining the concentration, order, separation, transport, and mixing of particles and liquids. However, the underlying physical principles guiding channel design are still limited, and most developments are based on empirical intuition. A deeper understanding of the basic physics and mechanisms of these systems will lead to more efficient and reliable designs with fewer iterations [38].

1.2.1 Reynolds Number

A wide range of physical phenomena occurs in microfluidic devices [37]. Dimensionless numbers are often used in microfluidics to determine the relative importance of different forces and interactions. These numbers represent the relative strength of various forces acting on a system [39]. For example, the Reynolds number (Re) is frequently employed to indicate the ratio of inertial to viscous forces. In most liquid microcurrents, the Reynolds number is often less than 10, indicating that viscous forces dominate fluid flow properties. Additionally, due to the low values of Reynolds number in microfluidics, diffusion plays a significant role in mass transfer, posing a potential challenge for mixing and exploiting micro-scale turbulence [39].

The Reynolds number (Re) is a widely used dimensionless quantity in microfluidics. Interestingly, despite microfluidic devices often operating in the low Reynolds number region to minimize inertia effects, it may be considered one of the less intriguing aspects of microfluidics. However, understanding microfluidics requires a strong intuition since low Reynolds number flows challenge our conventional understanding of fluid behavior. In his seminal work, Purcell's description of "life at low Reynolds number" (Purcell 1977) provides a comprehensive overview [37].

The Reynolds number of droplets is calculated based on the velocity during the pre-collision stage. It is defined as $Re = (\rho uL)/\mu$, where ρ is the droplet density (fluid density), u is the droplet velocity (fluid flow), L is the droplet characteristic size (a characteristic length), and μ is the fluid viscosity coefficient. In microchannels, the droplet velocity near the wall is slower than in the center, resulting in greater displacement of the droplet center compared to the peripheral fluid. Flows in microfluidics with $Re \leq 1$ are often considered almost Newtonian, with laminar flow and the absence of turbulence [40].

In microfluidics, when $Re \leq 1$, rapid mixing occurs mainly through diffusion, as turbulence is believed to be unlikely in low Reynolds number flows. Measuring turbulence in microchannels is challenging, and there is a lack of velocimeters capable of measuring it. A critical issue in microfluidics, due to laminar flow at low Reynolds numbers, is the relatively slow mixing of two fluids [40]. In their research, Wang et al. [40] made significant efforts to enhance mixing in microfluidics, employing high-voltage AC or DC to induce pressure in microchannels based on electrokinetic instability. However, these investigations focused on chaotic advection rather than turbulence. To understand turbulence, it is necessary to recognize its characteristics, including rapid propagation, random motion, high scattering velocity, continuous flow, multi-scale vortices, three-dimensionality, and high Reynolds numbers. In microfluidics, achieving turbulence requires high pressures [40]. Typically, there is no turbulence in microfluidics, and the flows remain laminar because the Reynolds number is ≤ 1. However, it can be observed that turbulence can be achieved at a low Reynolds number when the current is subjected to electrokinetic pressure in a T-like microchannel. A laser-induced fluorescence photobleaching anemometer was designed as a velocimeter with spatial resolution below the micrometer and microsecond timing to quantify high-frequency velocity oscillations in microchannels. Turbulence is usually considered a phenomenon occurring at high Reynolds numbers (Re), and it is widely accepted that turbulence does not exist in the flow of Newtonian fluids with low Re. Therefore, it is often assumed that flows in microfluidics can only be chaotic or laminar, even when electrokinetic forcing is used to increase flow instability. To understand the statistical characteristics and spatial patterns of multi-scale turbulent flows in microchannels, it is crucial to quantify the statistical properties of flow velocities with high spatial and temporal resolution [41].

According to the literature, microfluidic channels may not behave as theoretically expected when transitioning from laminar flow to turbulence. However, recent research suggests that the transition to turbulence in microchannels follows established laws, and any observed discrepancies may be due to experimental errors [32].

Low Reynolds number flows happen when fluid transfer systems are shrunk down to the micron and sub-micron size range. The way fluids move in these systems differs greatly from how fluids move on a large scale. Along with tiny data, we show visually how physical flows at low Reynolds numbers can change the design of a gadget. We also show an example of their work that uses fluid flow in tiny gadgets made for working with living things. The Navier–Stokes equation is the fundamental equation governing incompressible fluid dynamics:

$$\rho\left[\frac{\partial u}{\partial t}+(u.\nabla)u\right]=-\nabla p+\mu\nabla^2 u \tag{1.1}$$

The Navier–Stokes equation is an expression that describes the behavior of fluids and is analogous to Newton's law. It relates various parameters of the fluid, including the viscosity coefficient (μ), pressure (p), density (ρ), and velocity (u).

In the equation, the left side represents the mass-to-volume ratio (ρ) and the acceleration of the fluid. This formulation is based on the convective derivative in an Eulerian description. On the right side, some terms account for forces per unit volume. These include the pressure gradient ($-\ \nabla$P) and the viscous force term

($\mu\nabla^2 u$). Additionally, any additional body forces acting on the fluid, such as gravity (ρg), are also considered. The Reynolds number is a parameter without dimensions that aids in evaluating the equilibrium of inertial and viscous forces within a fluid. It is computed in the following manner:

$$\text{Re} = \rho \frac{\frac{u}{t} - u^2/l}{\mu u/l^2} \tag{1.2}$$

We have the following parameters: (ρ) for density, (u) for fluid velocity, (l) for a characteristic length scale (such as the diameter of a tube where velocity changes occur), (μ) for viscosity, and (t) for a characteristic time scale representing the time over which fluid velocity experiences significant changes in magnitude or direction. Considering the relationship (t X u ~ L), we can estimate Eq. (1.2) as:

$$\text{Re} = \frac{\rho u l}{\mu}$$

At low Reynolds numbers, fluid dynamics are primarily influenced by viscous forces rather than inertia. As a result, systems that rely on inertial forces for their efficiency become ineffective. This poses limitations when attempting to downscale a design that operates effectively at high Reynolds numbers. However, if a device is appropriately modeled, it can exhibit improved performance as its size decreases. When studying systems dominated by viscous forces, the approximations derived from such observations remain valid as long as the time scale, referred to as the "switching time" (T_s), and the Reynolds number are both less than 1. In this regime, the excitation pressure in the flow outweighs the characteristic time $\rho l^2/\mu$. It is important to note that as the diameter of the fluid channel decreases, these changes occur more rapidly [42].

1.2.2 Laminar Flow

As systems shrink to the micro-scale, the dominant forces governing physical phenomena undergo significant changes [43]. When macroscopic fluids are confined to small dimensions in microfluidic devices, the balance between inertia and viscosity is altered. Viscous forces become prominent in microfluidic channels, resulting in a reduced influence of inertia and mass transfer occurring primarily under the influence of viscosity [16]. By minimizing the impact of inertia, which is responsible for instability and turbulent behavior in liquids, a simplified flow regime known as laminar flow can be achieved [44]. Flow is typically categorized into two types: laminar flow and turbulent flow (Fig. 1.1) [42]. Laminar flow, characterized by smooth and predictable motion, allows for precise calculations of mass transfer over time, while turbulent flow exhibits erratic motion [43]. An example of laminar flow is the parallel

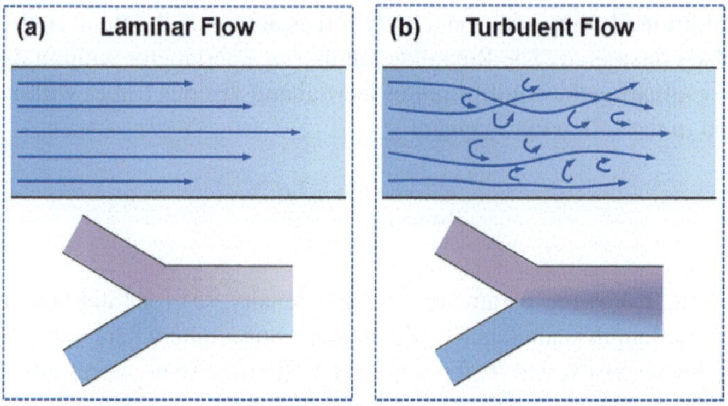

Fig. 1.1 The interfacial properties of liquid flow in microfluidic channels. **a** Laminar flow. At low Re (resistance less than 2000), the boundary between neighboring microfluidic flows is distinct and obvious. **b** Turbulent flow. At high Re, the microfluidic flow interface becomes less evident. Reprinted with permission from Ref [47]. Copyright 2022, Elsevier

flow of liquids [45]. The Reynolds number (Re) is used to describe the flow regime, indicating the relative importance of inertia compared to viscous stresses [46].

At low Reynolds numbers, inertia and turbulence have minimal effects, and the dominant factor is the sticky nature of fluid dynamics, resulting in laminar flow [44]. Due to the small size of micro-channels, flow is typically layered, and two or more currents cannot mix without diffusion. However, under certain circumstances, non-uniform penetration between the currents can occur due to variations in the microchannel height [32]. Flow in microfluidic channels generally occurs in the low Reynolds number regime, with at least one characteristic dimension in the micrometer scale and fluid velocities ranging from millimeters to centimeters per second ($\frac{cm}{s}/\frac{mm}{s}$). Laminar flow is observed at low Reynolds numbers ($Re < 1$), while higher Reynolds numbers ($Re > 2000$) often indicate turbulent flow [45].

The characteristics of laminar flow, such as its slow and controlled nature, make it possible to create unique testing platforms using microfluidics [43]. Experiments and particle sorting techniques have been developed by exploiting the interaction between stratified flows in microdevices. Laminar flow facilitates the controlled transfer of well-defined fluid packets on both sides, enabling precise cell analysis for various applications [32]. Additionally, laminar flow allows for seamless perfusion of different chemicals, making it suitable for localized cell therapy. By combining controlled diffusion and laminar flow, precise chemical gradients can be achieved, allowing for subcellular-level observation. However, constructing microfluidic pathways proves challenging due to their high surface-to-volume ratio, which can lead to the absorption of certain molecules into the device material [43].

1.2.3 Diffusion

Mixing involves the uniform distribution of all soluble molecules throughout the channel, and it can occur in two ways: through the transfer of liquids on a large scale or through diffusion, which dominates at the micron scale [42]. Diffusion is a crucial transport phenomenon that affects various processes [48]. Each process is associated with a specific time frame. While diffusion alone is impractical for mixing materials over a few centimeters, it is a fast and efficient process at the micron scale [42]. In macro-scale systems, random fluid vortices constantly stretch and mix the fluid, resulting in significantly reduced transport and mixing time scales. However, in microfluidic channels, convective mixing is absent due to the low Reynolds number, which is typically less than 1 [44]. Mixing in microchannels is solely achieved through molecular diffusion. Quantitative investigation of diffusion is required for nearly all processes in microfluidic devices. Accurate measurement of diffusion coefficients is a critical step in understanding and modeling transport phenomena in the selected system [48].

Using Brownian motion, a concentrated set of particles in a volume disperses over time, resulting in a constant average concentration throughout the volume [32]. Laminar flow in microfluidics allows different currents in contact with each other to mix through diffusion [46]. By stirring a binary mixture, two fluids can be dispersed throughout the receptacle on various scales. In such a scenario, all mixing occurs through diffusion, and the diffusion length required for thorough mixing is much smaller than the initial limit. If we ignore diffusion, we may express the time needed to stir two fluids into one by solving the equation $t_S = \frac{l_s}{v_s}$, where (v_s) is the typical stirring speed and (l_s) is the characteristic length. We can ignore diffusion in diameters above a few hundred microns since it is so sluggish. At low Reynolds numbers, any kind of agitation helps to mix because the fluid speeds induce shear and spread the fluid. A good stirring approach might be quite different from a bad one. A poor stirring approach is symmetrical over time, such as a piston that moves in and out of the fluid symmetrically. This method does not work well because, after one cycle, the liquid returns to its initial state (at Re = 0). On the other hand, an excellent stirring strategy is asymmetrical in time, such as a constantly rotating stirring rod. This approach ensures efficient mixing without any backtracking [14].

Regarding the fluid velocity profiles, electroosmotic fluid streams (a) and pressurized fluid streams (b) exhibit different profiles. The diffusion coefficient, $D = 10^{-5}$ ($cm^2\ s^{-1}$), for a globular particle (e.g., cell, molecule) can be determined using the following equation:

$$D = \frac{KT}{6\pi \mu r}$$

In the equation below, T represents temperature, K denotes Boltzmann's constant (1.38×10^{-23} J/K), r signifies the particle's hydrodynamic radius, and μ represents fluid viscosity. For a one-dimensional sample, x represents the distance a particle

Table 1.1 Exploring diffusion time in room temperature water [42]

Distance (μ)	Small molecule (s)	Protein (s)	Cell (s)
1	10^{-3}	10^{-2}	10^{1}
10	10^{-1}	10^{0}	10^{3}
100	10^{1}	10^{2}	10^{5}
1000	10^{3}	10^{4}	10^{7}

will move under the influence of diffusion in time t:

$$x = \sqrt{2Dt}$$

The diffusion coefficient is inversely proportional to the particle's hydrodynamic radius and shape. Small molecules have a greater diffusion coefficient and, on average, travel more per unit of time than large molecules do. Molecules of different sizes are separated using the change in diffusion coefficients over time [44]. Table 1.1 shows the time needed to diffuse at a specific distance. Mixing time scales change linearly with the length scale, while diffusion time is second-class. As a result, we understand that by decreasing the length scales (and the Reynolds number), we move from the state where diffusion time can be ignored relative to convection and mixing to a direction where diffusion can be very significant [42]. Since the distance with square power varies, small-scale diffusion becomes very important. For example, the hemoglobin with a speed of (7×10^{-7} cm^2 s^{-1}) in water takes 10^6 s to diffuse 1 cm, but only 1 s to diffuse 10 μm. However, due to diffusion, the distance traveled in a 10-micron-wide microchannel is critical. Microchannels can produce concentrate slopes with complicated profiles because diffusion periods are short at the microscale. To facilitate fast diffusion, small-scale mixing systems must discover strategies to maximize interfaces between solutions [32, 49].

Since diffusion occurs at such a small scale, small molecules from the fluid stream can be extracted quickly using this technique. The diffusion coefficient can be determined in microfluidic devices with a small sample volume (less than 1 ml) and a small amount of diffusing species (25–500 pmol). Results can be obtained in seconds or minutes with micro-devices, which is a short amount of time. For detecting the emission of microfluidic devices, so-called T-sensor and H-cell micro-devices, which function continuously, have garnered a lot of attention in recent years [48].

1.2.4 T-Sensor

Diffusion is important in the T-sensor, which is used as a diagnostic tool for chemical tests [50–52]. The T-sensor induces signal changes by facilitating diffusion between the analyte and the indicator through the convergence of two or more input currents, typically associated with a physical parameter such as analyte concentration (refer to Fig. 1.2) [53]. Operating at low Reynolds numbers, the T-sensor ensures laminar flow.

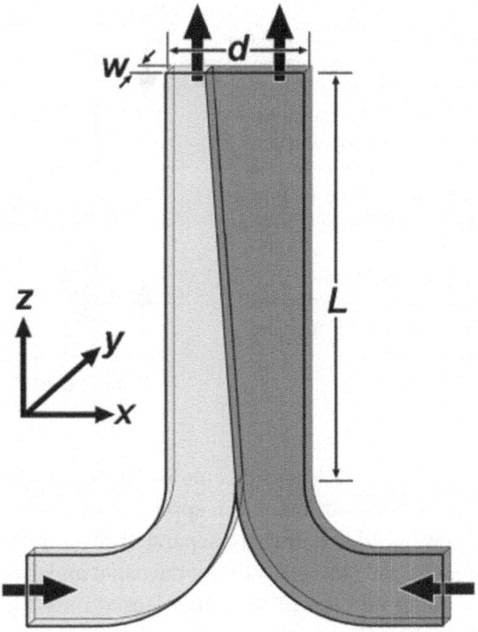

Fig. 1.2 This conceptual illustration showcases the T-sensor design, featuring two fluid inlets at the bottom. The fluid on the right contains a diffusible analyte (black) that diffuses along the length of the channel. During operation, the optical signal is measured downstream of L following the substantial diffusion process. Reprinted with permission from Ref. [53]

Direct visualization of molecule diffusion between parallel flowing streams is possible. The propagation mode is influenced by the pressure velocity characteristics of the duct apparatus, which impacts the utilization of the T-sensor for measuring molecular attributes [53, 54]. T-sensors are utilized in various biochemical assays, including diffusion measurement and reaction kinetics, competitive safety analysis, and analyte concentration determination. By adjusting the flow velocities, reactions occurring at the interface can be precisely controlled, allowing for precise product management. The size of the T-sensor determines molecule diffusion, enabling manipulation of the interaction between antibodies and antigens. Concentrations of dilute analytes can be determined by imaging the steady state of designated components in a flowing stream, requiring only a small sample volume of a few microliters. High-speed assays using hormones, pharmaceuticals, and other small analytes achieve detection limits ranging from 1 to 10 nm [44].

1.2.5 H-Filter

Figure 1.3 illustrates the schematic of the H filter [55]. Two parallel currents flow through a channel and enter the system through separate ports. When the channel length or sample flow is sufficiently large, important particles diffuse and fill the tube before exiting, resulting in particle separation. However, particles with lower diffusion factors remain in the half-channel and are not captured. H filters have been

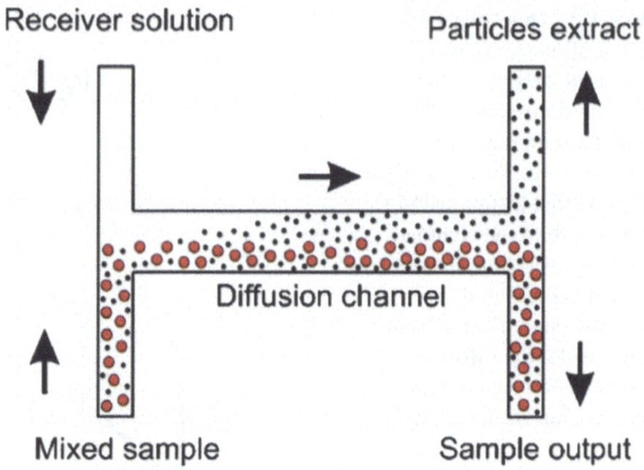

Fig. 1.3 An H filter is denoted, wherein the length l is chosen in such a manner that large particles are unable to scatter within the channel and consequently remain restricted within their main flow; simultaneously, smaller intended molecules permeate through the pathway in the neighboring stream. Reprinted with permission from Ref. [55]. Copyright 2011, Elsevier

used for size-based molecule isolation, distinguishing motile sperm from immotile sperm, and various other applications [44]. A specific type of H filter is shown where motile sperms are separated from immobile ones [56]. Motile sperms swim rapidly and randomly, filling the canal and displaying different dispersion compared to nonmotile sperm that are released only through diffusion. Accordingly, rapidly spreading species are extracted. The device, designed by Schilling et al. [57], can perform analysis based on these principles. In the H-filter, when a flow of cells comes into contact with a chemical factor, the cell membrane lyses, allowing certain intracellular chemicals to be released. The output of the H filter consists of intracellular molecules and waste cells [37].

1.2.6 Surface Area to Volume Ratio

The surface-to-volume ratio of a system increases as its size decreases, especially in microdevices where the ratio can reach magnitudes on the order of 10^6 m. At such small scales, inertial forces become less significant compared to viscous, electrostatic/electrodynamic, or surface tensile forces. Surface forces that may not be noticeable at macroscopic sizes play a critical role in the functioning of microfluidic devices [39].

For instance, a vessel with a diameter of 35 mm and a volume of 2.5 mL has a surface area to volume ratio (SAV) of 4.2 cm^{-1}. In contrast, a microchannel with dimensions of 50 μm in height, 30 mm in length, 50 μm in width, and a capacity of 75 nL has a surface area to volume ratio of 800 cm^{-1} [32].

Microfluidic devices, characterized by narrow cross-sections, exhibit an exceptionally high surface-to-volume ratio. This attribute makes microchannels well-suited for capillary electrophoresis (CE) applications as they efficiently dissipate excess heat from the sample. However, a major drawback of the high surface-to-volume ratio is channel wall deposition, where macromolecules and cells can adhere to the micro-walls, potentially reducing electroosmotic pumping efficiency, causing blockages, and affecting system performance [44].

1.2.7 Surface Tension

At the micrometer scale, surface tension becomes a significant force [42]. Surface tension arises from the cohesion between liquid molecules at the interface between liquid and gas [32]. Laplace's law can be used to calculate the equivalent pressure due to surface tension [42].

$$\Delta \mathrm{P} = \gamma \left(\frac{1}{R_1} + \frac{1}{R_2} \right)$$

The pressure generated at the surface of a liquid can be determined by considering the vertical radii of curvature, R_1 and R_2. The surface free energy of the liquid is denoted by γ. When the length, R, representing the curvature of the wall approaches infinity, the equation can be simplified [32].

$$\Delta \mathrm{P} = \frac{\gamma}{R}$$

This equation represents the pressure at the fluid boundary between two large parallel planes separated by a distance of 2R. If R_2 equals R_1 and the surface is spherical, the equation can be further simplified [8].

$$\Delta \mathrm{P} = \frac{2\gamma}{R}$$

Using this approach, the pressure within a spherical liquid droplet can be calculated. This concept applies generally [42].

$$\mathrm{P} = \frac{2\gamma \mathrm{COS}\theta}{\mathrm{r}}$$

The contact angle between the surface and the liquid, as well as the radius of curvature of the interface, determines the force needed to overcome surface tension. In the case of a hydrophilic device, such as one treated to attract water, the liquid will enter the channel but come to a stop at the exit. A pressure of at least 2 cos/r (where r is the channel radius) is necessary to overcome the surface tension and produce a continuous flow. The pressure produced in a 3-μm diameter pipe is almost equal to one atmosphere, a significant value, when the contact angle (θ) is equal to 1, signifying a hydrophilic surface. The surface tension-related pressure that needs to be overcome is detailed in Table 1.2. In the case of a hydrophobic surface, the necessary pressure is identical to that which is necessary to force water into the channel [42].

Table 1.2 The influence of surface tension on water pressure within pipes [42]

Radius (μ)	Pressure (at 1 atm)
0.15	10^7
1.5	10^6
15	10^4
150	10^3
1500	10^2

Temperature, electrostatic potential, and surfactant concentration have significant effects on the interfacial tension between fluids. By manipulating these properties, gradients in surface tension can be created externally [42]. Surface energy plays a crucial role in facilitating fluid transfer within a system. When the surface is hydrophobic, the absorption of energy allows water to penetrate without the need for pressure or specific gap sizes [58]. These forces can be employed to control fluid flows. For instance, plants with large leaves experience substantial water evaporation, resulting in significant water release. However, this process can pose challenges in fluid systems due to the presence of tiny air bubbles. Overcoming these bubbles requires applying high pressure, creating a potential issue in fluid systems. Air bubbles can form in two ways: in hydrophilic systems, air can get trapped where a larger channel joins a smaller one, due to the thin wetting layer that forms across the smaller outlet before the air can pass through. Alternatively, air bubbles can appear when the device becomes wet, and air spontaneously escapes from the solution. This occurrence is often observed when a cold solution is heated, reducing the solubility of gases in the liquid.

Water exhibits the highest surface tension among common liquids, with nearly three times the surface tension compared to other fluids. To mitigate the presence of air bubbles during wetting, one effective approach is to pre-wet the channel with a liquid such as isopropyl alcohol, which has lower surface tension than water. Subsequently, water can be introduced without the formation of air–water interfaces. This method reduces the surface tension force by up to three times, enabling the removal of air bubbles through constrictions. For further details, refer to Table 1.3 [42]. The surface energy of a fluid directly relates to its surface tension and influences the height of the water capillary. In microfluidics, surface forces play a vital role in creating virtual walls and pumping mechanisms, especially in channels with micron dimensions [32].

Table 1.3 Scaling behavior of parameters for the microfluidic device at the constant time [42]

Volume	L^3
Reynolds number	L^2
Length	L^1
The pressure of surface tension	L^{-1}
Diffusion time	L^{-2}

1.3 Manufacturing Methods

Microfluidic technology offers a broad range of miniaturized analytical devices, commonly associated with lab-on-chip (LOC) systems or micro-total-analysis systems (μTAS). These systems perform scaled-down operations on miniaturized models of conventional laboratory benchtop devices [59]. The selection of appropriate materials for microfluidic device fabrication is a fundamental aspect of microfluidic applications [60]. Given the amplification of attributes on a microscale surface, the choice of the primary material is expected to impact the properties of the resulting nanomaterials. Capillary microfluidics exhibit unique phenomena such as large surface-to-volume ratios, enhanced mass and heat transfer, laminar flows, and reduced maintenance times. Unlike macro-scale channels, the wetting and contact angles of anhydrous solutions on chip surfaces play a critical role [61]. When selecting a material, factors such as surface functionalization, pressure, and temperature requirements, chemical compatibility with the intended reagents, biocompatibility, stability, transparency, ease of manufacturing, and overall chemical and physical properties need to be considered [7].

Microfluidic devices and systems are made from a wide range of materials. Because of its well-established fine machining capabilities, silicon is one of the most often utilized materials in micro/nanosystems. The ability to incorporate control/sensing circuits into a semiconductor, robust manufacturing procedures, good chemical resistance, and desirable mechanical qualities are all benefits of employing silicon as a substrate or structural material. Other materials used for substrates and micro/nanostructures include polymers, metals, ceramics, quartz, and glass, depending on the application. Plastic substrates, particularly polymers, have gained significant attention and are considered promising substrates due to their optimal biochemical reliability and compatibility, cost-effectiveness, and ease of manufacturing. Plastics such as polycarbonate, polyethylene, poly(dimethylsiloxane) (PDMS), polyamide, and poly(methyl methacrylate) (PMMA) offer a wide range of chemical and physical material parameters suitable for laboratory chip applications, often manufactured using low-cost replication methods. Plastics and polymers are preferred materials in microfluidic and laboratory chip applications, allowing for techniques like injection molding, hot embossing, and mass production through casting. The primary challenge in microfluidic system fabrication typically lies in the formation of microfluidic channels, which serve as crucial microstructures/nanostructures within the lab-on-chip [59]. Common methods for constructing microfluidic devices include micromachining, embossing, soft lithography, in situ manufacturing, laser abrasion, and injection molding. Each method has its benefits and limitations, and the choice of fabrication method often hinges on the device's intended use [32].

1.3.1 Micromachining

Microfluidics emerged alongside the development of the microelectronics industry and silicon-based microelectromechanical systems (MEMS) to meet the biological (chemical) requirements of micro-total-analysis systems (μTAS). Consequently, advanced micro-manufacturing techniques were promptly employed to fabricate microchannels for liquid and gas chromatography and capillary electrophoresis. The ability to perform microanalysis with high resolution and sensitivity became a significant driving force behind early microfluidics. Initially, first-generation microfluidic chips were developed and manufactured using traditional machining methods on silicon/glass substrates [15].

Silicon micromachining, a well-established technique in microelectromechanical systems, played a prominent role in the early stages of microfluidics [5]. This method enabled the fabrication of complex systems using silicon as the primary material. Recent advancements in nanotechnology have further allowed for the creation of nanoscale structures for microfluidic systems [32]. Silicon has been a preferred choice for constructing microfluidic systems due to its widespread availability, thermal stability, and chemical compatibility. Its ease of fabrication, design flexibility, potential for surface modifications, and semiconductor properties have made silicon the go-to material for microfluidic devices over the years [29]. However, while widely used, silicon is no longer the preferred material for contemporary microfluidic applications, and several limitations must be acknowledged when using it in practical settings. One notable drawback is silicon's opacity, rendering it unsuitable for light detection in the visible and ultraviolet regions. If on-site imaging is required, a non-silicon component must be incorporated. Additionally, the integration of active components, such as pumps and valves, becomes challenging in silicon-based systems due to their fragility and high elastic modulus. Moreover, the cost of silicon is not particularly advantageous, making the material relatively expensive. Nevertheless, silicon-based microfluidic systems find utility in biological applications, such as organ-on-chip devices and medical diagnostics, where they are employed for drug toxicity screening [29].

Glass-based micro-reactors offer a multitude of desirable properties that make them well-suited for chemical reactions under extreme conditions, such as high pressures, high temperatures, and aggressive solvents. Glass is chemically inert, heat-resistant, biologically compatible, rigid, electrically insulating, and easily modifiable on the surface. These characteristics enable glass micro-reactors to perform well in challenging environments. With higher micrometer-scale resolution compared to other materials, glass microcapillary reactors are ideal for precise control over the production of polymer emulsions and nanoparticles [62].

Glass possesses several advantages for microfluidic devices. It can incorporate active components, provides excellent optical transparency, and is more cost-effective than silicone. Glass also finds widespread use in biological and chemical laboratories due to its compatibility with biological models and overall popularity. Therefore, it comes as no surprise that glass is a popular choice for microfluidic chips. Typically, soda-lime glass, borosilicate glass, and molten quartz are used as substrates in the construction of glass microfluidic devices [63].

However, glass-based microfluidic devices face certain limitations in terms of manufacturing challenges. While glass itself is relatively inexpensive, the production of glass chips requires clean room facilities, which adds to the overall cost. The process is time-consuming and demands specialized skills and equipment. Despite these limitations, glass remains indispensable for specific applications that necessitate its unique properties. For instance, when handling strong solvents, maintaining stable chemical surfaces, or operating at high temperatures, glass or silicon-based microfluidic devices are preferred. Glass machining continues to be the primary choice for chip-based capillary electrophoresis due to the surface properties of glass.

The traditional micromachining process consists of three steps. The initial stage is lithography, where the substrate is first cleaned and then coated with a photoresist layer (PR). Next, a mask with transparent and opaque zones is used to cover the PR layer, and selective removal of the PR allows the geometric micro-pattern on the mask to be transferred to the layer. The next step is etching, where a specific etchant is used to selectively remove the substrate and form microchannels with the desired shape. Finally, bonding is the critical third step for creating closed channels. The two layers of the micromachine are aligned and adhered to each other, either through thermic fusion or adhesive bonding, to create a sealed microfluidic chip with closed channels. Glass and silicone offer distinct advantages in specific situations. Their thermal stability makes them suitable for high-temperature chemical syntheses, while their solvent resistance allows for various droplet-based effects in natural environments. Additionally, their unique optical properties, stable electroosmotic current, and high thermal conductivity make them essential in droplet microfluidics. However, their impermeability and hardness can limit their use under certain conditions, and in some biomedical analyses conducted in an aqueous environment, glass or silicone may not be necessary.

Furthermore, the high cost of large-scale manufacturing, the requirement for cleanroom equipment, lengthy manufacturing processes, and overall expenses associated with cleanroom facilities pose challenges in the widespread adoption of glass and silicone microfluidic devices [15]. PDMS has been the standard material for building microfluidic devices in the previous decade. It can be cast into intricate molds and tightly bonded to glass. The elastomeric nature of PDMS allows for easy integration of fluid valves and pumps on-chip, facilitating the fabrication of multilayer devices. However, PDMS has drawbacks compared to amorphous silicon and glass thermoplastics due to its low hardness (generally less than 1 MPa). This characteristic makes it difficult to create high aspect-ratio channels as PDMS tends to collapse. Moreover, the automation of soft material usage, like PDMS, has posed challenges, impeding efforts to boost PDMS production. While the high permeability of PDMS to oxygen and water can be advantageous in certain applications, it also presents problems in various scenarios. The hydrophobic characteristic of PDMS benefits certain biological applications, allowing water-soluble medicines and metabolites like urea or albumin to penetrate the material of drug screening devices through hydrophobic-hydrophobic interactions. However, due to their low reproducibility, these materials are better suited for research laboratories than for industrial applications. As alternatives for microfluidic elastomers, extra-stoichiometric thiol-one (OSTE) and thermoplastic elastomers offer more desirable features such as hardness, permeability, and the potential for rapid processing [64].

1.3.2 Soft Lithography

The controlled manipulation of liquids on sub-millimeter scales offers numerous advantages for chemical and biological applications. These advantages include biochemical diffusion times and fast thermal, laminar flow profiles, a small volume of reactants, and precise handling of fine particles such as microorganisms and living cells.

To leverage these benefits, researchers have employed conventional micromachining methods to construct microfluidic devices on a small scale. Initially, these methods involved using traditional semiconductor materials like glass and silicon. However, the drawbacks associated with these materials prompted the development of exemplary laboratory methods for microfluidic device fabrication [65].

The rapid progress in microfluidics has increased the demand for rapid prototyping and mass production of chips for both industrial and laboratory research purposes. Furthermore, the growing utilization of clinical and biomedical microfluidic chips requires cost-effective and disposable solutions to prevent contamination. Therefore, the polymer model molding approach, particularly through droplet microfluidic chips, has gained recognition. One widely employed method for replicating molds is soft lithography, as depicted in Fig. 1.4 [66]. Soft lithography enables the creation of multiple transparent polymer chips from a single mold. In this process, a two-part polymer called polydimethylsiloxane (PDMS), consisting of an elastomer and a curing agent, is molded with photo-resistance masters [32]. To initiate the process, a PDMS replica or mold is created by mixing the base and curing agent components. The silicon hydride groups in the curing agent react with the vinyl groups in the base, resulting in the formation of a cross-linked elastomeric solid [67].

Soft lithography is a widely used technique in microfabrication that involves casting polydimethylsiloxane (PDMS) in a photo-resistant mold, typically made of SU-8. In some cases, patterned silicon wafers are used as molds when high aspect ratio features are desired due to the bending properties of photo-resist materials. PDMS offers several advantages over silicon, including its softness, gas permeability, biocompatibility, and transparency, making it well-suited for various biological applications [32].

To create a replica, the two components of PDMS, typically mixed in a base-to-curing agent ratio of 10:1 (v/v), are blended. The resultant liquid pre-polymer is poured over the main liquid and cured. This approach enables the PDMS pre-polymer to adhere to the geometry of the original master, faithfully duplicating its features with a precision as low as 10 nm. PDMS's low surface-free energy and resilience enable easy separation from the masters without causing damage. However,

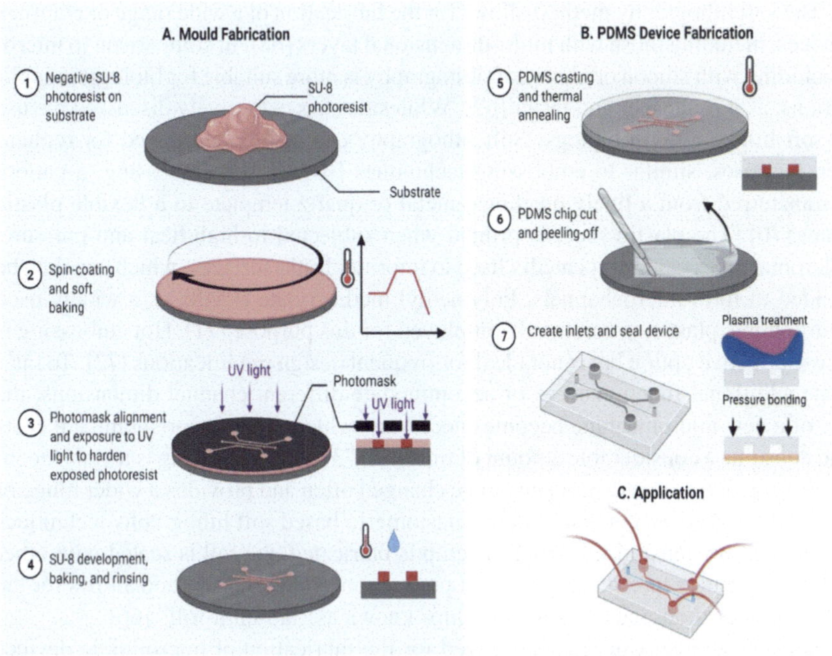

Fig. 1.4 **a** Fabrication of an SU-8 master mould; **b** casting of polydimethylsiloxane (PDMS) on the mould, plasma treatment, and contact pressure bonding; **c** integrating PDMS microfluidic device with world to chip interfaces for application. Reprinted with permission from Ref. [5]. Copyright 2021, MDPI

the production of the master mold is often the most challenging aspect of replica PDMS manufacturing, and various techniques can be employed to obtain a high-quality master. One common method involves using high-resolution transparency as a photomask for photolithography [67].

In the initial stage of the process, a two-dimensional photomask is created using computer-assisted design software, specifying the desired channel geometry. This design is then transferred onto a silicon wafer coated with a photosensitive polymer through exposure to ultraviolet light. The polymer undergoes polymerization in areas where light passes through the photomask, resulting in patterns corresponding to the designed channels. After development, only these patterned areas remain, serving as molds into which clear PDMS will be poured. Subsequent freezing and baking steps are performed to remove the polymer and create openings for sample injection through the channels. In the final stage, the PDMS replica is bonded to a glass surface or other transparent substrates, creating a sealed chip ready for testing. It's important to note that while soft lithography is suitable for rapid prototyping, it is not yet optimized for low-cost mass production, as it can be time-consuming and expensive, taking several hours to produce a single chip [66].

The soft lithography method allows for the fabrication of a wide range of complex devices, including those with multi-dimensional layers [68]. In comparison to micromachining with silicon or glass, soft lithography is more suitable for biological applications, as it is cheaper and faster [62]. Whitesides has extensively discussed the use of soft lithography in biology. Soft lithography can also be employed for replication purposes, similar to embossing techniques [69]. In hot embossing, a pattern is transferred from a finely machined metal or quartz template to a flexible plastic sheet [70]. The plastic sheet is printed when subjected to high heat and pressure. Micromachines can be repeatedly used to imprint plastic surfaces, which can then be bonded to form microchannels. Polymethyl methacrylate (PMMA), a widely used hydrophobic plastic, is commonly employed for this purpose [71]. Hot embossing is a cost-effective option but is not ideal for frequent design modifications [72]. To integrate additional functionalities or accommodate different channel dimensions, the use of a new micromachine becomes necessary, which might incur significant costs and consume a considerable amount of time [72, 73]. Hot embossing is a viable technique for patterns that do not need to be changed often and provides a wider range of material alternatives compared to the elastomeric-based soft lithography techniques mentioned previously [32]. After the chip is fabricated, the foil is sealed with other polymer layers before starting the next production cycle. This method allows for the production of thousands of low-cost chips known as "lab-on-a-foil" [66].

Soft lithography can also be utilized for the fabrication of microfluidic devices through a technique known as "multilayer soft lithography." This approach involves the integration of multiple interconnected substrates, allowing each layer to serve a distinct function. By combining different substrates, it is possible to incorporate both rigid and flexible components within the device. This enables the routing of fluids while accommodating electronic circuitry. Additionally, the use of multiple substrates allows for the incorporation of diverse mechanical and physical functionalities. Furthermore, each substrate can be engineered with specific biochemical features, facilitating the creation of multilayer microfluidic systems that incorporate permeable substrates for enhanced biochemical transport in biological investigations [65].

1.3.3 Molding

Numerous replicas can be made from a single original using replica molding, which is a technique for creating PDMS structures from a silicon master. This method can be performed in a regular laboratory environment without the need for a costly clean room. It allows for the reproduction of certain properties as small as 10 nm in size. Replica molding is a part of the broader category of soft lithography techniques, which involve manipulating elastomeric structures through methods such as micro-contact printing, embossing, injection molding, and casting [45].

Among these techniques, injection molding is a particularly promising and cost-effective approach for manufacturing microfluidic devices. It makes use of soft, flexible thermoplastic polymeric materials that have been heated to the glass transition temperature. Next, a mold cavity is filled with the melted plastic. The plastic cools quickly in the cavity because its temperature is lower than the plastic, and the molded component is ready in a matter of minutes. The time-consuming stage in this process is building the master mold that shapes the plastic. Precision molding tools, also known as masters, can be created using various techniques such as silicon micromachining, electroplating, and metal micromachining. Injection molding and hot embossing share many similarities in terms of molding procedures and cost considerations for the master mold. However, injection molding is preferred for mass manufacturing due to its cost-effectiveness and shorter processing time compared to hot embossing. In microfluidic injection molding, resolution and material selection are critical factors that require careful consideration [8]. Another method for fabricating microfluidic devices is casting materials onto a pre-patterned mold, which is a form of micro-molding. This technique is an extension of using silicone stamps (PDMS) to cast heat-resistant or UV polymers. It is particularly suitable for transforming liquid-state biological materials into solid masses. Hydrogels, both synthetic and natural, are commonly used biomaterials in micro-molding. Hydrogels have the advantage of natural porosity and can mimic the extracellular matrix, making them suitable as scaffolds for engineering living tissues. The casting method is often employed in producing microfluidic biomaterials and shares similarities with the conventional PDMS-based methods used in soft lithography. These techniques are relatively easy to learn and not technically difficult. While micro-molding can be automated, it is commonly used for manual production of a small number of specimens. It is also well-suited for creating scaffolds for micro-physiological systems with relatively simple grid geometries [74]. By utilizing replica molding, injection molding, and casting techniques, researchers and manufacturers can fabricate microfluidic devices with high precision and cost-effectiveness, enabling a wide range of applications in fields such as biology, chemistry, and medical diagnostics.

1.3.4 Other Methods

1.3.4.1 3D Printed Transfer Molding

Microfluidic chips can now be manufactured using 3D printing technology, which offers several advantages in the fabrication process. 3D printing makes it possible to create 3D structures in a single step by using computer-controlled layers of materials to form the desired object. The geometric information of the structure is encoded in CAD files, which are utilized along with computer-controlled printers. This enables the realization of the intended structural design through "mail-order" services, resulting in more accurate builds with lower labor costs, shorter production times, and higher resolution. Additionally, 3D printing is environmentally friendly

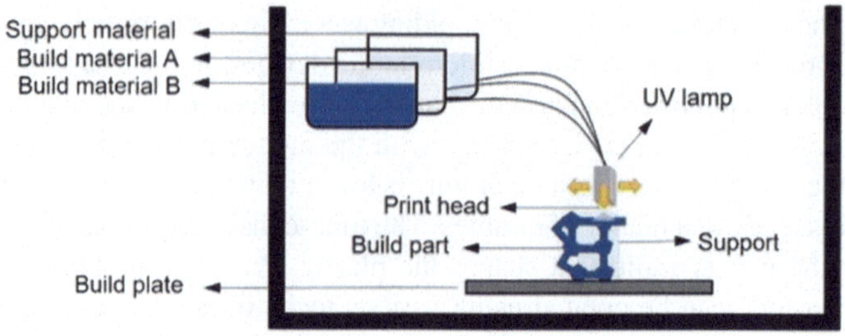

Fig. 1.5 A 3D-printed item is made utilizing MJM. Photoreactive liquid resin in droplet form is expelled from a nozzle onto a construction platform and quickly cured by UV radiation. At the same time, the support material is released from a separate nozzle, and the platform falls as the 3D-printed item is made. Reprinted with permission from Ref. [75]. Copyright 2023, MDPI

as it is an additive manufacturing technology that eliminates the need for removing excess material. Moreover, by combining industrial-grade user-embedded control systems and interfaces, this fabrication technique can be commercialized. These characteristics indicate a potential trend in droplet microfluidic chip production [15].

One prototype of this technology for microfluidic applications involves the formation of elastomeric polymers using 3D-printed molds. McDonald et al. utilized the previous "solid object printing" method, also known as MJM, to 3D print various molds for PDMS micro-molding. These molds were used to control microfluidic systems with spiral channels and multiple layers. As MJM is shown in Fig. 1.5.

Researchers have expanded the application of 3D printing technology to create templates using SLA (Stereolithography) and MJM (Multi-Jet Modeling). They have also utilized 3D printing to create internal templates that can be removed after preparing the PDMS (Polydimethylsiloxane) material. For example, Huang et al. developed MJM 3D printing molds that allowed the liquid PDMS to fill complex channel architectures. After the PDMS was cured, the molds were removed, resulting in intricate PDMS channel structures (Fig. 1.6).

Similarly, Costa et al. [77] employed a three-step process using 3D printing molds to create a "thrombosis chip." They first used SLA to print microchannel patterns that simulated healthy and diseased vascular structures. Then, PDMS was cast over the printed patterns, and finally, the 3D printed patterns were removed, leaving microchannels embedded in solid PDMS.

Glick et al. [78] proposed a double-sided molding technique that enabled the one-step production of thin membranes. Using 3D printing to create molds offers several advantages: each mold can be used to produce multiple microfluidic devices, saving time and reducing the cost of 3D printing. Additionally, this approach is compatible with common microfluidic materials such as PDMS, allowing researchers to employ established molding techniques and achieve complex geometric features that are difficult to obtain using conventional methods. However, it is important to note that molds have limitations in terms of geometric diversity since they need to facilitate the molding process. While 3D printing transfer molding techniques can overcome

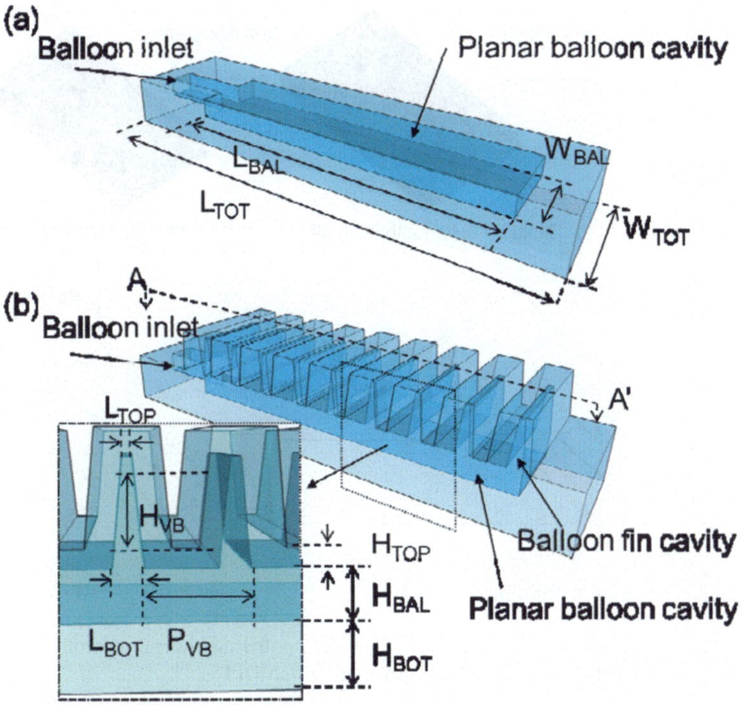

Fig. 1.6 Pattern for 3D printing to create 3D molds for PDMS processing. Illustrations of molding PDMS using a distinct 3D printing mold, which can then be detached. Reprinted with permission from Ref. [76]. Copyright 2015, Elsevier

challenges associated with processing wafers in clean room environments, there are still many wafer molding protocols that need to be considered after fabrication [66].

Amidst the various types of 3D printing processes, stereolithography (SLA) stands out as a highly effective method for creating microfluidic chips. SLA involves the polymerization of photocurable resin liquids, substrate by substrate, using laser radiation (as illustrated in Fig. 1.7). In the production of microfluidic chips, the cavity walls of the channels are polymerized, while the unpolymerized resin is subsequently removed.

Bertsch et al. [80] were among the first to utilize SLA for manufacturing microfluidic chips, benefiting from its cost-effectiveness and time efficiency. Importantly, SLA allows for the creation of complex channels and intricate structures that may be challenging to achieve using other methods. More recently, Thompson et al. [81] employed the SLA process to produce a library of standardized parts and fittings for use in droplet emulsion circuits. Unlike other modular technologies, SLA can be easily reconfigured and self-aligned, allowing for the development of hierarchical 3D objects. The 2-photon polymerization technique, when used to SLA, may absorb two photons, increasing clarity beyond what would be possible with optical diffraction alone. In addition, SLA encompasses a wide range of microfluidic 3D printing

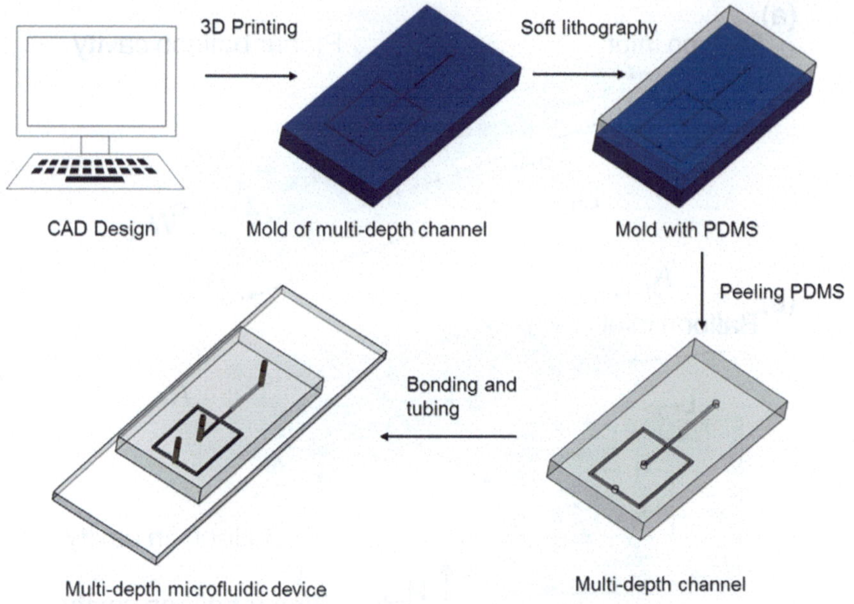

Fig. 1.7 The production process of an LCD SLA 3D printed multi-depth microfluidic device. Reprinted with permission from Ref. [79]. Copyright 2019, MDPI

processes. The process of obtaining the final microfluidic design can be quite laborious. Initial testing of microfluidic devices often reveals design and operational flaws, necessitating design modifications. These iterations can lead to significant delays and increased development time, consequently driving up costs. However, with the advent of 3D printing technology, it is now possible to fabricate entire microfluidic devices in a single step, eliminating the need for assembly, such as with PDMS. Recently, there have been advancements in the 3D printing of microfluidic devices incorporating membrane-based valves. By utilizing the appropriate material formulation and a 3D printing machine, researchers were able to create the first functional microfluidic device with an integrated valve in just one hour. This significant reduction in design time showcases the potential of 3D printing to accelerate the development process. It transforms the landscape of microfluidic device development by enabling a "fast and fail" strategy, where initial prototypes can be rapidly iterated based on experimental feedback.

In the following sections, we provide a more detailed description of various 3D printing technologies and their applications in the field of biological microfluidics. The ability to use CAD models for microfluidic device design allows for the integration of commercially available components with known or measurable dimensions, facilitating chip connectivity. For instance, Jida and colleagues researchers demonstrated the ease with which multiple electrochemical sensing schemes can be integrated into 3D-printed microfluidic chips. They showcased different versions of microfluidic devices with applications in NO detection, neurotransmitter detection, and oxygen pressure measurement in red blood cells. The designs also incorporated fluid connections and membrane inserts for detecting signaling chemicals.

Unlike soft lithographic masters, part files for 3D printing can be standardized and easily shared across laboratories equipped with CAD applications and 3D printers capable of producing various medical components [82–84]. This portability enables immediate experimentation after exposing a biological sample to the device, akin to a portable experimental model [85]. In the development of Microfluidics Interface (MFI) technology, which aims to improve existing processors, 3D printing is also helpful [86]. MFI makes it possible to integrate on-chip devices with a variety of materials and capabilities in a seamless manner [87]. Researchers led by Hwanyong Lee were able to make a polymer MFI using SLA that works well for an advanced on-chip integrated reverse transcription (RT) microchip that can do both RNA extraction and cDNA synthesis at the same time.

1.3.4.2 Paper-Based Microfluidic Chips

Paper is well acknowledged as an easily manipulable substrate in the production of microfluidic devices [88–90]. Paper-based microfluidic chips are typically created using embossing, photolithography, and capillary wetting techniques [91]. These processes allow for easy preparation, surface-related reactions, and pump-free operation [92]. Additionally, three-dimensional microfluidic chips can be fabricated using origami techniques [93]. Furthermore, a bio-printing methodology has been developed for controlled deposition and arrangement of hydrogel fiber models, enabling the assembly of microfluidic devices and microchannel networks [94–96]. This method offers an accurate three-dimensional environment for studying cells and has a lot of promise for tissue engineering and organ-on-a-chip applications (Fig. 1.8). Different manufacturing techniques may combine diverse materials with diverse chemical and physical properties into a single chip. The creation of hybrid glass-PDMS microdevices is one well-known example. The glass substrate reduces the interactions between fluids and elastomers, which lets you fine-tune the chemistry of the channel surface. The PDMS elastomer makes it easy to connect pneumatic pumps and valves. Additionally, the concept of "co-fabrication" has been employed to produce organized, multifunctional microchannels. Through the use of multi-groove PDMS molds, channels can exhibit heterogeneous spatial functions by selectively removing additives. Moreover, microfluidic chips can incorporate components such as optical elements and microheaters, enabling multiple functionalities. With ongoing multidisciplinary research and technological advancements, the manufacturing of microfluidic chips will expand beyond the described technologies, incorporating a wider range of materials and applications [15].

Recently, a cheaper and simpler method for producing microfluidic chips has emerged, allowing for the fabrication of paper chips. In this approach, the paper is impregnated with an optical insulating material before specific patterns are created using ultraviolet radiation [98]. The polymerized regions act as hydrophobic barriers

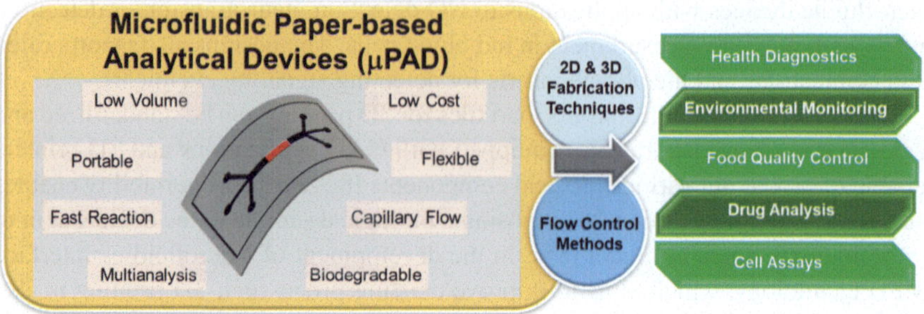

Fig. 1.8 Diagrammatic representation illustrating the benefits and potential uses of 2D and 3D microfluidic paper-based analytical devices. Reprinted with permission from Ref. [97]. Copyright 2019, MDPI

for the paper channels, with the black areas of the photomask corresponding to the desired tube locations. Simultaneously, other photoresist residues are removed during development. This significantly simplified fabrication technique reduces the cost per chip by up to 1%. Alternatively, hydrophobic PDMS or wax can be directly printed onto the paper layer using modified inkjet printers. Olkonen et al. proposed a method for creating hydrophobic barriers by printing polystyrene flexographically to form loop-to-loop structures. Paper chips do not require external pumps to drive fluid flow in the channels; instead, samples are easily transported through capillary forces exerted within the three-dimensional matrix of the paper. This principle has been well validated. Furthermore, samples can be transferred from one area of the chip to another (rather than simply spreading over a wide area) by simply pulling buffer liquid from a reservoir above the sample surface. Volumetric flows can also be controlled by introducing "capillary pumps," such as a thick piece of paper acting as a sponge, at various distances from the sample. The longer the distance, the slower the flow rate of a particular sample, which can be easily determined by the increased resistance along the channel(s) (linear scale of resistance with a given channel length) [66]. Microfluidic paper-based analyzers (μPADs) have gained attention in the development of point-of-care testing (POCT) due to their potential as single-use, user-friendly, and integrated diagnostic platforms [99–101].

The manufacturing of μPADs offers several advantages [102–106]:

1. Cost-effectiveness, time efficiency, portability, lightweight design, low power consumption (eliminating the need for pumps or external equipment), durability, and suitability for low-resource environments worldwide.
2. No requirement for cleanroom facilities.
3. Minimal liquid volume (microliters) needed for analysis.
4. Simultaneous generation of multiple evaluations.
5. Compatibility with telemedicine, enabling healthcare services in resource-constrained settings.

However, μPADs also have limitations [107–111]:

1. Sample retention and evaporation can hinder efficient sample delivery during transportation.
2. Current approaches for whole-genome/transcriptome amplification before sequencing may suffer from bias and nonspecific products.
3. Complex processes in the analysis, such as material mixing prior to the final reaction, may not be suitable for home-based healthcare.

Given that research on μPADs is still in its early stages, substantial efforts are required to make them practical and truly beneficial, particularly in expanding their applications in POCT [101]. Future development of μPADs should focus on studying printing technology and paper structure to reduce evaporation and sample retention [110]. Achieving precise sample flow control through considerations such as paper surface energy and vertical flow in addition to capillary compaction warrants further investigation. Switches and paper power supplies are also vital parameters that can attract more research. Our ongoing research aims to enhance μPAD detection by improving specificity, selectivity, sensitivity, and ease of operation. Exploring bioactive paper through the integration of biofunctional and novel nanomaterials presents promising avenues to boost the selectivity and sensitivity of diagnostic instruments. Molecular detection represents a distinct diagnostic method to enhance the specificity and sensitivity of μPADs. Although Rodriguez et al. [112] established lateral molecular detection experiments with paper-based chromatography, these results provide a baseline for future developments in molecular detection based on μPADs. We think that future research and initiatives will primarily concentrate on merging μPADs with nucleic acid extraction, diagnostic results, and isothermal amplification to overcome the obstacles mentioned [88].

However, each method to fabricate a chip, has different advantages and disadvantages. Table 1.4 shows the advantages and disadvantages of each method.

1.4 Conclusion

In conclusion, this chapter has provided a concise survey of the valuable and eloquent uses of microsystem technologies, focusing on the physics of microfluidic devices and their manufacturing methods. The primary physical effects governing microfluidic behavior have been explored, shedding light on the unique characteristics exhibited at the microscale. It has been emphasized that microfluidic systems should not be treated as scaled-down versions of macroscopic systems but rather as distinct entities that require a ground-up approach in their design and fabrication. The importance of understanding fluid dynamics in microchannels has been highlighted, with an emphasis on low flow rates, the absence of turbulence, and the dominance of diffusion for mixing. Furthermore, the stronger surface tension forces at the air/liquid interface in microfluidic systems have been acknowledged as a critical factor to consider. Regarding manufacturing methods, the chapter has touched upon both traditional and non-traditional techniques for creating microfluidic devices from materials such as

Table 1.4 A comparison of different methods to prepare chips and biochips

Method	Disadvantages	Advantages
Micromachining	• Difficulties in assembling components, high cost, optical turbidity, and surface properties make silicon unsuitable for biological applications • Limited scalability for mass production • Time-consuming process • Requires specialized skills, facilities, and equipment	• High precision and resolution • Enables complex and intricate channel designs • Compatible with a wide range of materials, including silicon and glass • Suitable for chemical applications requiring stable chemical surfaces, high temperatures, or strong solvents
Soft lithography	• Limited resolution compared to micromachining • Material compatibility and bonding challenges • Not suitable for cheap mass production • Requires access to cleanroom facilities for photolithography	• Cheaper, faster, and more convenient than micromachining for biological applications • Allows for rapid prototyping and iterative design changes • Suitable for a variety of microstructures
Molding	• Requires a master mold fabrication step • Limited to simpler channel geometries • Challenges with demolding and air bubble entrapment • Material selection and resolution limitations	• Cost-effective for large-scale production • Good replication fidelity • Suitable for creating elastomeric structures like PDMS
3D printing	• Limited resolution compared to other techniques • Material selection and compatibility constraints • Post-processing steps may be required for surface treatment or sealing • Mold adaptability is limited to support the molding process • Some limitations in achieving certain geometric properties	• Enables rapid prototyping and customization • Complex designs, including embedded features and multi-material structures, can be achieved • On-demand manufacturing capability • Compatibility with common microfluidic materials, particularly PDMS • Time and cost savings through the use of related micro-molding procedures • Allows for the production of intricate geometries that are difficult to achieve with conventional methods

(continued)

Table 1.4 (continued)

Method	Disadvantages	Advantages
Paper base	• Inefficient sample delivery due to retention and evaporation during transportation • Limitations for complex analysis processes, such as sample mixing • Potential bias and nonspecific products in genome/transcriptome amplification • Limited to two-dimensional designs • Material constraints and potential limitations in chemical compatibility	• Low-cost, time-efficient, and low consumption • Strong and suitable for diagnostic tools in resource-constrained environments • No cleanroom facilities required • Requires microliter volume of liquids • Multiple assays can be generated simultaneously • Can be used with telemedicine to provide healthcare services

silicon, glass, and plastic. The choice of fabrication method and materials depends on the specific application and desired functionalities of the microfluidic device. Overall, this chapter has provided valuable insights into the physics of microfluidics and their manufacturing methods, paving the way for further advancements in this rapidly evolving field. By harnessing the unique physical phenomena at the microscale and employing appropriate fabrication techniques, microfluidic systems hold great promise for enabling innovative applications that would otherwise not be possible.

References

1. Niculescu, A.-G., et al.: Fabrication and applications of microfluidic devices: a review. Int. J. Mol. Sci. **22** (2021). https://doi.org/10.3390/ijms22042011
2. Bragheri, F., Martínez Vázquez, R., Osellame, R.: Chapter 12.3—microfluidics. In: Baldacchini, T. (ed.) Three-Dimensional Microfabrication Using Two-Photon Polymerization, second edn., pp. 493–526. William Andrew Publishing (2020)
3. Shanehband, N., Naghib, S.M.: Microfluidics-assisted tumor cell separation approaches for clinical applications: an overview on emerging devices. Comb. Chem. High Throughput Screen (2024)
4. Sharan, P., et al.: Microfluidics for microswimmers: engineering novel swimmers and constructing swimming lanes on the microscale, a tutorial review. Small **17**(26), 2007403 (2021)
5. Scott, S.M., Ali, Z.: Fabrication methods for microfluidic devices: an overview. Micromachines **12**(3), 319 (2021)
6. Shanehband, N., Naghib, S.M.: Recent advances in nano/microfluidics-based cell isolation techniques for cancer diagnosis and treatments. Biochimie **220**, 122–143 (2024)
7. Garshasbi, H.R., Naghib, S.M.: A review on nano/microfluidic devices for cell isolation techniques: recent progress and advances. Curr. Nanosci. **20**(6), 717–739 (2024)
8. Naghib, S.M., Mohammad-Jafari, K.: Microfluidics-mediated liposomal nanoparticles for cancer therapy: recent developments on advanced devices and technologies. Curr. Top. Med. Chem. (2024)

9. Mohammad-Jafari, K., Naghib, S.M.: 3D printing of microfluidic-assisted liposomes production for drug delivery and nanobiomedicine: a review. Curr. Med. Chem. (2024)
10. Sartipzadeh, O., et al.: Microfluidic-assisted synthesis and modeling of stimuli-responsive monodispersed Chitosan microgels for drug delivery applications. Sci. Rep. **12**(1), 8382 (2022)
11. Sartipzadeh, O., et al.: Microfluidic-assisted synthesis and modelling of monodispersed magnetic nanocomposites for biomedical applications. Nanotechnol. Rev. **9**(1), 1397–1407 (2020)
12. Rahmanian, M., et al.: A micropillar array-based microfluidic chip for label-free separation of circulating tumor cells: the best micropillar geometry? J. Adv. Res. **47**, 105–121 (2023)
13. Sartipzadeh, O., et al.: Controllable size and form of droplets in microfluidic-assisted devices: effects of channel geometry and fluid velocity on droplet size. Mater. Sci. Eng. C **109**, 110606 (2020)
14. Shi, W., et al.: Integrated silicon microfluidic chip for picoliter-scale analyte segmentation and microscale printing for mass spectrometry imaging. Lab Chip **23**(1), 72–80 (2023)
15. Shang, L., Cheng, Y., Zhao, Y.: Emerging droplet microfluidics. Chem. Rev. **117**(12), 7964–8040 (2017)
16. Yang, R.-J., et al.: Micro-magnetofluidics in microfluidic systems: a review. Sens. Actuators B Chem. **224**, 1–15 (2016)
17. Xiao, Z., et al.: A silicon-based fuel cell micro power system using a microfabrication technique. J. Micromech. Microeng. **16**(10), 2014 (2006)
18. Xu, C., Cai, L., Huang, Z.: Ultra-microscale chemistry experiment on microfluidic devices: a new trend in chemical education. J. Chem. Educ. **101**(4), 1564–1573 (2024)
19. Aubry, G., Lee, H.J., Lu, H.: Advances in microfluidics: technical innovations and applications in diagnostics and therapeutics. Anal. Chem. **95**(1), 444–467 (2023)
20. Stone, H.A., Kim, S.: Microfluidics: basic issues, applications, and challenges. Am. Inst. Chem. Eng. AIChE J. **47**(6), 1250 (2001)
21. Tapre, R.W., et al.: Microscale fluid mechanics: a review of material-dependent flow characteristics and applications. Mater. Today Proc. (2023)
22. Gravesen, P., et al.: Microfluidics—a review. J. Micromech. Microeng. **3**(4), 168 (1993)
23. McDonald, J.C., Duffy, D.C., Anderson, J.R., Chiu, D.T., Wu, H., Schueller, O.J.A., Whitesides, G.M.: Fabrication of microfluidic systems in poly(dimethylsiloxane). Electrophoresis, **21**(1), 27–40 (2000)
24. Quake, S.R., Scherer, A.: From micro- to nanofabrication with soft materials. Science, **290**(5496), 1536–1540 (2000)
25. Wu, J., et al.: Modular microfluidics for life sciences. J. Nanobiotechnol. **21**(1), 85 (2023)
26. Kumari, M., et al.: Microfluidics-based nanobiosensors for healthcare monitoring. Mol. Biotechnol. **66**(3), 378–401 (2024)
27. Guo, J., et al.: Multiple bio-actives loaded Gellan Gum microfibers from microfluidics for wound healing. Small **19**(44), 2303887 (2023)
28. Behera, P.P., et al.: Integrated microfluidic devices for point-of-care detection of bio-analytes and disease. Sens. Diagn. (2023)
29. Niculescu, A.-G., et al.: Fabrication and applications of microfluidic devices: a review. Int. J. Mol. Sci. **22**(4), 2011 (2021)
30. Cardoso, B.D., et al.: Recent advances on cell culture platforms for in vitro drug screening and cell therapies: From conventional to microfluidic strategies. Adv. Healthcare Mater. **12**(18), 2202936 (2023)
31. Yang, W., Hou, L., Luo, C.: When Super-Resolution Microscopy Meets Microfluidics: Enhanced Biological Imaging and Analysis with Unprecedented Resolution. Small **19**(23), 2207341 (2023)
32. Beebe, D.J., Mensing, G.A., Walker, G.M.: Physics and applications of microfluidics in biology. Annu. Rev. Biomed. Eng. **4**(1), 261–286 (2002)
33. Tabeling, P.: Recent progress in the physics of microfluidics and related biotechnological applications. Curr. Opin. Biotechnol. **25**, 129–134 (2014)

34. Tabeling, P.: Introduction to Microfluidics. Oxford University Press (2023)
35. Gil, J.F., et al.: Cancer models on chip: paving the way to large-scale trial applications. Adv. Mater. **35**(35), 2300692 (2023)
36. Hupp, A.M., Kovarik, M.L., McCurry, D.A.: Emerging areas in undergraduate analytical chemistry education: microfluidics, microcontrollers, and chemometrics. Annu. Rev. Anal. Chem. **17** (2024)
37. Squires, T.M., Quake, S.R.: Microfluidics: fluid physics at the nanoliter scale. Rev. Mod. Phys. **77**(3), 977 (2005)
38. Amini, H., Lee, W., Di Carlo, D.: Inertial microfluidic physics. Lab Chip **14**(15), 2739–2761 (2014)
39. Chakraborty, D., Chakraborty, S.: Microfluidic transport and micro-scale flow physics: an overview. Microfluid. Microfab. 1–85 (2010)
40. Wang, G., Yang, F., Zhao, W.: There can be turbulence in microfluidics at low Reynolds number. Phys. Rev. E **14**(8), 1452–1458 (2014)
41. Wang, G., Yang, F., Zhao, W.: Microelectrokinetic turbulence in microfluidics at low Reynolds number. Phys. Rev. E **93**(1), 013106 (2016)
42. Brody, J.P., et al.: Biotechnology at low Reynolds numbers. Biophys. J. **71**(6), 3430–3441 (1996)
43. Breslauer, D.N., Lee, P.J., Lee, L.P.: Microfluidics-based systems biology. Mol. Biosyst. **2**(2), 97–112 (2006)
44. Holmes, D., Gawad, S.J.: The application of microfluidics in biology. Microeng. Biotechnol. 55–80 (2010)
45. Sia, S.K., Whitesides, G.M.: Microfluidic devices fabricated in poly (dimethylsiloxane) for biological studies. Electrophoresis **24**(21), 3563–3576 (2003)
46. Chen, J., et al.: Progress of microfluidics for biology and medicine. Nano-Micro Lett. **5**, 66–80 (2013)
47. Cheng, S., Chen, W., Zhang, P.J.G.: Developing advanced polymer films based on microfluidic laminar flow. Giant **9**, 100091 (2022)
48. Miložič, N., et al.: Evaluation of diffusion coefficient determination using a microfluidic device. Chem. Biochem. Eng. Quart. **28**(2), 215–223 (2014)
49. Aryasomayajula, A., et al.: Microfluidic Devices and Their Applications, pp. 487–536 (2017)
50. Brody, J.P., Yager, P., Goldstein, R.E., Austin, R.H. Biotechnology at low Reynolds numbers. Biophys. J., **71**(6), 3430–3441 (1996)
51. Kamholz, A.E., Weigl, B.H., Finlayson, B.A., Yager, P.: Quantitative analysis of molecular interaction in a microfluidic channel: the T-sensor. Anal. Chem., **71**(23), 5340–5347 (1999)
52. Weigl, B.H., Yager, P. Microfluidic diffusion-based separation and detection. Science, **283**(5400), 346–347 (1999)
53. Kamholz, A.E., Yager, P.: Theoretical analysis of molecular diffusion in pressure-driven laminar flow in microfluidic channels. Biophys. J. **80**(1), 155–160 (2001)
54. Baroud, C.N., et al.: Reaction-diffusion dynamics: Confrontation between theory and experiment in a microfluidic reactor. Phys. Rev. E **67**(6), 060104 (2003)
55. Shafique, M.F., et al.: Fabrication of embedded microfluidic channels in low temperature co-fired ceramic technology using laser machining and progressive lamination. Eur. Ceram. Soc. **31**(13), 2199–2204 (2011)
56. Cho, B.S., Schuster, T.G., Zhu, X., Chang, D., Smith, G.D., Takayama, S. Passively driven integrated microfluidic system for separation of motile sperm. Anal. Chem., **75**(7), 1671–1675 (2003)
57. Schilling, E.A., Kamholz, A.E., Yager, P.: Cell lysis and protein extraction in a microfluidic device with detection by a fluorogenic enzyme assay. Anal. Chem., **74**(8), 1798–1804 (2002)
58. Hossain, A., et al.: Engineering bioinspired microfluidics; biomimetic self-healing/cleaning coating designs and unique advanced materials. Chem. Eng. J. 151–336 (2024)
59. Lei, K.F.: Materials and Fabrication Techniques for Nano-and Microfluidic Devices (2014)
60. Gimondi, S., et al.: Microfluidic devices: a tool for nanoparticle synthesis and performance evaluation. ACS Nano **17**(15), 14205–14228 (2023)

61. Grimmer, A., Wille, R.: Designing Droplet Microfluidic Networks. Springer (2020)
62. Ma, L., et al.: Droplet microfluidic devices: working principles, fabrication methods, and scale-up applications. Small Methods **8**, 2301406 (2024)
63. Sebastian, V.: Toward continuous production of high-quality nanomaterials using microfluidics: nanoengineering the shape, structure and chemical composition. Nanoscale **14**(12), 4411–4447 (2022)
64. Iliescu, C., et al.: A practical guide for the fabrication of microfluidic devices using glass and silicon. Biomicrofluidics **6**(1) (2012)
65. Sochol, R.D., et al.: 3D printed microfluidics and microelectronics. Microelectron. Eng. **189**, 52–68 (2018)
66. Eicher, D., Merten, C.A.: Microfluidic devices for diagnostic applications. Exp. Rev. Mol. Diagnost. **11**(5), 505–519 (2011)
67. McDonald, J.C., Whitesides, G.M.: Poly (dimethylsiloxane) as a material for fabricating microfluidic devices. Acc. Chem. Res. **35**(7), 491–499 (2002)
68. Huang, Z., Shao, G., Li, L.: Micro/nano functional devices fabricated by additive manufacturing. Prog. Mater. Sci. **131**, 101020 (2023)
69. Juska, V.B., et al.: Silicon microfabrication technologies for biology integrated advance devices and interfaces. Biosens. Bioelectron. **237**, 115503 (2023)
70. Lin, X., et al.: Fabricating biomimetic materials with ice-templating for biomedical applications. Smart Med. **2**(3), e20230017 (2023)
71. Rahul, R., et al.: A mould-free soft-lithography approach for rapid, low-cost and bulk fabrication of microfluidic chips using photopolymer sheets. Microfluid. Nanofluid. **27**(11), 78 (2023)
72. Xie, M., et al.: Functional microfluidics: theory, microfabrication, and applications. Int. J. Extreme Manufact. **6**(3), 032005 (2024)
73. Ece, E., Hacıosmanoğlu, N., Inci, F.: Microfluidics as a ray of hope for microplastic pollution. Biosensors **13**(3), 332 (2023)
74. Tien, J., Dance, Y.W.: Microfluidic biomaterials. Adv. Healthc. Mater. **10**(4), 2001028 (2021)
75. Sommonte, F., Denora, N., Lamprou, D.A.J.P.: Combining 3D printing and microfluidic techniques: a powerful synergy for nanomedicine. Pharmaceuticals **16**(1), 69 (2023)
76. Hwang, Y., et al.: Pneumatic microfinger with balloon fins for linear motion using 3D printed molds. Sens. Actuators A Phys. **234**, 65–71 (2015)
77. Costa, P.F., Albers, H.J., Linssen, J.E.A., Middelkamp, H.H.T., van der Hout, L., Passier, R. et al.: Mimicking arterial thrombosis in a 3D-printed microfluidic in vitro vascular model based on computed tomography angiography data. Lab Chip. **17**, 2785–2792 (2017)
78. Glick, C.C., Srimongkol, M.T., Schwartz, A.J., Zhuang, W.S., Lin, J.C., Warren, R.H. et al.: Rapid assembly of multilayer microfluidic structures via 3D-printed transfer molding and bonding, Microsyst. Nanoeng. **2**, 16063 (2016)
79. Mohamed, M.G., et al.: Rapid and inexpensive fabrication of multi-depth microfluidic device using high-resolution LCD stereolithographic 3D printing. J. Manufact. Mater. Process. **3**(1), 26 (2019)
80. Serex, L., Bertsch, A., Renaud, P.: Microfluidics: A new layer of control for extrusion-based 3D printing. Micromachines **9**(2) 86 (2018)
81. Thompson, B., Movsesian, N., Cheng, C., Karandikar, P., Gupta, M., Malmstadt, N.: Modular microfluidics for double emulsion formation. Methods in Cell Biology, **148**, 161–176 (2018)
82. Roopavath, U.K., Kalaskar, D.M.: Introduction to three-dimensional printing in medicine. In: 3D Printing in Medicine, pp. 1–27. Elsevier (2023)
83. Nazir, A., et al.: Multi-material additive manufacturing: A systematic review of design, properties, applications, challenges, and 3D printing of materials and cellular metamaterials. Mater. Des. **226**, 111661 (2023)
84. Ramasamy, M., et al.: Inexpensive and rapid fabrication of PDMS microfluidic devices for biological testing applications using low cost commercially available 3D printers. J. Micromech. Microeng. **33**(10), 105016 (2023)

85. Saha, R., et al.: Evolution of 3d printing technology in fabrication of microfluidic devices and biological applications: a comprehensive review. J. Micromanufact. **7**, 25165984241237356 (2024)
86. Milton, L.A., et al.: Vat photopolymerization 3D printed microfluidic devices for organ-on-a-chip applications. Lab Chip **23**(16), 3537–3560 (2023)
87. Hagemann, C., et al.: Low-cost, versatile, and highly reproducible microfabrication pipeline to generate 3D-printed customised cell culture devices with complex designs. PLoS Biol. **22**(3), e3002503 (2024)
88. Xia, Y., et al.: Fabrication techniques for microfluidic paper-based analytical devices and their applications for biological testing: a review. Biosens. Bioelectron. **77**, 774–789 (2016)
89. Li, M., et al.: On-chip liquid manipulation via a flexible dual-layered channel possessing hydrophilic/hydrophobic dichotomy. ACS Appl. Mater. Interfaces **15**(15), 19773–19782 (2023)
90. Yedire, S.G., et al.: Additive manufacturing leveraged microfluidic setup for sample to answer colorimetric detection of pathogens. Lab Chip **23**(18), 4134–4145 (2023)
91. Marangon, F., Baumgartner, D., Planchette, C.: In-air microfluidic strategy for the production of sodium alginate fibers with regular inclusions at very high throughput. Phys. Rev. Appl. **19**(5), 054006 (2023)
92. Tycova, A., Prikryl, J., Hemzal, D.: Capillary electrophoresis and Raman: can we ever expect light at the end of the tunnel? TrAC Trends Anal. Chem. **161**, 117017 (2023)
93. Li, B., et al.: A brief overview of passive microvalves in microfluidics: mechanism, manufacturing, and applications. Biomicrofluidics **18**(2) (2024)
94. Tong, X., et al.: A novel colorimetric and fluorescent dual-mode 3D paper-based microfluidic aptasensor for the point-of-care detection of sulfadimethoxine. Sens. Actuators B Chem. **397**, 134691 (2023)
95. Wang, X., et al.: Electric yo-yo centrifugation combining with paper-based microfluidic immunoassay chip for inflammatory biomarkers detection in whole blood. Talanta **253**, 123883 (2023)
96. Zhang, Y., et al.: Pb (II) inhibits CRISPR/Cas12a activation and application for paper-based microfluidic biosensor assisted by smartphone. Sens. Actuators B Chem. **398**, 134732 (2024)
97. Lim, H., Jafry, A.T., Lee, J.J.M.: Fabrication, flow control, and applications of microfluidic paper-based analytical devices. Molecules **24**(16), 2869 (2019)
98. Dwivedi, S.: Fabrication techniques and materials for bio-MEMS. In: MEMS and Microfluidics in Healthcare: Devices and Applications Perspectives, pp. 101–141. Springer (2023)
99. Bilgen, E., et al.: Disposable paper-based sensors. In: Fundamentals of Sensor Technology, pp. 803–860. Elsevier (2023)
100. Rypar, T., et al.: Low-tech vs. high-tech approaches in μPADs as a result of contrasting needs and capabilities of developed and developing countries focusing on diagnostics and point-of-care testing. Talanta **266**, 124911 (2023)
101. Rink, S., Baeumner, A.J.: Progression of paper-based point-of-care testing toward being an indispensable diagnostic tool in future healthcare. Anal. Chem. **95**(3), 1785–1793 (2023)
102. Yamaguchi, H., Miyazaki, M.: Enzyme-immobilized microfluidic devices for biomolecule detection. TrAC Trends Anal. Chem. **159**, 116908 (2023)
103. Chauhan, P.S., Sharma, N., Tomar, R.S.: Sensing technologies used in microfluidic paper-based analytical devices (μPADs). In: Paper-Based Diagnostic Devices for Infectious Diseases, pp. 1–12. IOP Publishing Bristol, UK (2023)
104. Faheem, A., Cinti, S.: Advanced techniques for manufacturing paper-based microfluidic analytical devices. In: Microfluidic Biosensors, pp. 159–170. Elsevier (2023)
105. Moulahoum, H.: Dual chromatic laser-printed microfluidic paper-based analytical device (μPAD) for the detection of atrazine in water. ACS Omega **8**(44), 41194–41203 (2023)
106. Shalaby, A.A., et al.: Microfluidic paper-based analytical devices for cancer diagnosis. Sens. Actuators B Chem. **379**, 133243 (2023)
107. Silva-Neto, H.A., et al.: Recent advances on paper-based microfluidic devices for bioanalysis. TrAC Trends Anal. Chem. **158**, 116893 (2023)

108. Li, W., et al.: Review of paper-based microfluidic analytical devices for in-field testing of pathogens. Anal. Chim. Acta **1278**, 341614 (2023)
109. Brazaca, L.C., et al.: The use of biological fluids in microfluidic paper-based analytical devices (μPADs): recent advances, challenges and future perspectives. Biosens. Bioelectron. **246**, 115846 (2023)
110. Kulkarni, M.B., et al.: Recent advances in microfluidics-based paper analytical devices (μPADs) for biochemical sensors: from fabrication to detection techniques. Biochem. Eng. J. **198**, 109027 (2023)
111. Alahmad, W., Varanusupakul, P., Varanusupakul, P.: Recent developments and applications of microfluidic paper-based analytical devices for the detection of biological and chemical hazards in foods: a critical review. Crit. Rev. Anal. Chem. **53**(2), 233–252 (2023)
112. Rodriguez, N.M., Linnes, J.C., Fan, A., Ellenson, C.K., Pollock, N.R., Klapperich, C.M.: Paper-Based RNA Extraction, In Situ Isothermal Amplification, and Lateral Flow Detection for Low-Cost, Rapid Diagnosis of Influenza A (H1N1) from Clinical Specimens. Anal. Chem., **87**(15), 7872–7879 (2015)

Chapter 2
Exploring Microelectromechanical Systems (MEMS) and Their Manufacturing Processes

Abstract This chapter presents a comprehensive overview of microfabrication techniques and materials utilized in the manufacturing of microelectromechanical systems (MEMS). MEMS technology has emerged as a pioneering frontier in the realm of miniaturization, offering immense potential across various industries. The chapter commences by introducing MEMS and highlighting its pivotal role in facilitating the miniaturization of intricate devices. Subsequently, it delves into the fundamental processes involved in MEMS fabrication, encompassing thin-film deposition, lithography, wet and dry etching, and substrate bonding. Furthermore, the chapter explores the burgeoning field of soft micromachining, emphasizing the unique advantages conferred by soft materials like polymers and gels. Overall, this chapter provides a comprehensive understanding of the crucial microfabrication processes and materials underpinning MEMS production, ultimately paving the way for future advancements in this rapidly evolving field.

Keywords Microelectromechanical systems (MEMS) · Manufacturing processes · Technology · Miniaturization · Fabrication

2.1 Introduction

Microelectromechanical systems (MEMS) are integrated microdevices that combine mechanical and electrical components [1, 2]. These systems can actuate, control, sense, and function on a microscale, either independently or in arrays, to achieve macro-scale efficiencies [3]. Since the mid-1970s, MEMS technology has gained recognition for its potential in actuator applications and the development of sensors for biological, chemical, and physical parameters [4]. MEMS devices are manufactured using batch processing methods that are compatible with Integrated Circuit (IC) fabrication techniques [5]. While MEMS technology is derived from IC production processes, the testing methodologies of the two technologies are significantly different [6]. The concept of miniaturization and creating small machines was discussed by physicist Richard Feynman in 1959 [7], laying the foundation

S. M. Naghib and Y. Rezaian, *Frontiers of Nanobiotechnology*, Series in BioEngineering, https://doi.org/10.1007/978-3-031-88419-1_2

for the development of MEMS technology. In the 1970s, researchers at Stanford University, including Kurt Petersen [8] and Richard Muller, paved the way by exploring the miniaturization of mechanical systems using silicon-based fabrication processes. Their groundbreaking work demonstrated the feasibility of creating intricate microscale structures integrated with electronics, propelling the advancement of MEMS technology [9–11].

MEMS devices, which respond to various inputs such as electrical, biological, optical, physical, and chemical, can now be constructed at sizes too small to be seen by the naked eye, typically measured in micrometers or even microns [12]. The miniaturization of mechanical systems offers unique opportunities for scientific and technological progress, enabling the creation of smaller, lighter, faster, and more accurate devices compared to their macro mechanical counterparts [12]. These devices can be seamlessly integrated with electronics to form high-performance closed-loop microelectromechanical systems, utilizing microfabrication techniques commonly employed in IC manufacturing [13]. Throughout the 1980s and 1990s, MEMS technology progressed rapidly, with researchers and engineers developing various fabrication techniques like surface and bulk micromachining to create complex microstructures and components [14]. The integration of MEMS with electronics became a major focus, leading to the development of sensors, actuators, and systems on a chip [15–17]. The commercial applications of MEMS have expanded significantly over the years, finding use in diverse industries [18–20]. Inkjet printers utilize MEMS technology for precise ink deposition, while automotive applications rely on MEMS-based sensors for airbag deployment and stability control [21]. Consumer electronics incorporate MEMS devices in displays, and the healthcare sector benefits from MEMS pressure sensors and medical diagnostics [22, 23]. MEMS technology has also opened doors to innovative applications in areas such as microsurgery, DNA analysis, and personalized medicine [24].

2.2 Applications and Impact of MEMS Technology in Different Fields

Advancements in manufacturing technology have brought about a remarkable breakthrough: the ability to build actuators and sensors on the same microscopic plane as a processor chip [25–28]. This revolutionary approach allows for integrating motors, micro-mirrors, pumps, valves, and various sensors, such as flow, viscosity, inertial, pressure, and thermal sensors, onto a single chip alongside the processor unit [29–33]. These components, measured in microns, come together to form a compact system on a chip [34]. This remarkable progress in technology has opened up a world of possibilities with limitless potential [35, 36]. One exciting application is the development of medical and biological devices that can be miniaturized to a degree where they can be injected directly into the bloodstream [37]. These tiny devices can selectively target and destroy diseased cells or bacteria while leaving healthy tissue unharmed [38].

Moreover, they can intelligently monitor blood substances and release medicines when required. In the field of microsurgery, intelligent MEMS devices have revolutionized procedures by enabling surgery to be performed inside the human body without the need for incisions [39]. There is even the potential for on-site DNA analysis and sequencing using these advanced devices [40–42].

The progress in microelectromechanical systems (MEMS), microsystems technology, and micromachining has also facilitated the creation of small biomedical devices for a range of applications. These include lab-on-a-chip systems, neural prosthetics implants, and drug delivery systems. Compared to devices manufactured using traditional methods, these technologies offer numerous advantages. They allow for lower costs through batch fabrication, reduced dimensions, integration of sensing and actuating functions on the same layer, improved functionality, and the ability to enhance reliability through redundancy using multiple small devices. Glass and silicon have conventionally been the primary materials employed in micromachining and MEMS [43]. Microsystems, as a field of study, focuses on systems at the microscale. Researchers in this field explore materials and techniques that enable the creation of tiny devices with characteristics comparable to larger-scale systems but with distinct features and behaviors. It is worth noting that microfabrication methods can be synergistically combined with industry-standard procedures to generate improved or even novel products [44].

The commercial applications of MEMS technology are driving the demand for MEMS devices in the future. Various examples illustrate this trend, including inkjet printers utilizing piezoelectrics or thermal bubble ejection for precise ink deposition, acceleration sensors in automobiles for airbag deployment, incorporation of MEMS devices in consumer electronics like gaming controllers, MP3 players, and digital cameras, as well as other applications such as hard disk head parking to prevent data loss, utilization of MEMS gyroscopes for stability control systems, silicon pressure sensors in blood pressure monitors and automobile tire pressure sensors, micromirrors used in displays such as DMD chips in DLP projectors, optical switching techniques for data communications, applications of interferometric modulator displays (IMOD) in consumer electronics, and the use of RF MEMS technology to enhance the performance of inductors and capacitors [45, 46]. MEMS technology has paved the way for the development of an extensive range of optical, radiofrequency, and nanodevices. MEMS, or Microelectromechanical Systems, are known by various names based on their constituent parts and intended uses. In radio-frequency applications, they are called radio-frequency MEMS or RF. For optical applications, they are referred to as micro opto-electro-mechanical systems (MOEMS). When at least one component has a size smaller than 1 μm, they are classified as nanoelectromechanical systems (NEMS). In cases where MEMS incorporate biological materials, such as DNA strands, to detect targets or influence cells, they are known as bioMEMS [47].

2.2.1 Microfabrication Techniques

Microfabrication techniques are essential for the effective utilization of sophisticated and cost-effective microfluidic devices [48, 49]. The choice of suitable materials, such as paper, polymers, silica, metals, and silicon, is crucial in the microfabrication process [50]. These techniques have evolved from the Micro-Electro-Mechanical Systems (MEMS) technologies that gained popularity alongside semiconductor technology [51]. The foundation of microfabrication techniques can be traced back to the development of integrated circuits in the 1950s. The discovery of the piezoresistive phenomenon in germanium (Ge) and silicon (Si) paved the way for the utilization of these materials, which offered significantly higher instrument sensitivity compared to metal films. The miniaturization of integrated circuitry accelerated over the years, driven by Moore's Law, which observed a doubling of transistors in a dense integrated circuit every two years. This pattern continues to hold, with IC chips now featuring nanoscale features and the scaling process ongoing. MEMS devices benefit from the same microfabrication techniques that have enabled the miniaturization of integrated circuits [52].

The microfabrication process for MEMS devices relies on several fundamental manufacturing processes that have been used by microchip makers for almost 40 years. These processes include substrate bonding, etching, lithography, and thin-film deposition. Thin films are deposited using various physical or chemical methods to perform structural tasks and provide isolation. Lithography transfers the desired pattern onto the thin film, similar to traditional photography and printing methods. Etching, achieved through a chemical solution, is used to selectively remove parts of the substrate. Finally, layer bonding is employed to combine multiple functionalities or achieve packaging goals, and these stages can be repeated as needed [53]. These microfabrication techniques have revolutionized the manufacturing of MEMS devices, allowing for the creation of intricate structures and integrated functionalities. The ongoing advancements in microfabrication continue to push the boundaries of what is possible in terms of miniaturization and device performance.

2.3 Overview of Microfabrication Techniques for MEMS Production

Microelectromechanical systems (MEMS) fabrication processes encompass a variety of techniques and methods used to create intricate and functional devices at a microscale level. These processes can be broadly categorized into two groups: hard techniques and soft techniques [54]. Hard techniques in MEMS fabrication involve well-established and traditional processes that draw inspiration from the semiconductor industry and integrated circuit manufacturing. They include thin-film deposition, lithography, etching (both wet and dry), substrate bonding, and various forms of micromachining such as bulk, surface, and high-aspect-ratio micromachining [55, 56].

On the other hand, soft techniques have emerged as an alternative approach to MEMS fabrication. These techniques utilize soft materials such as elastomers or polymers to create flexible and biocompatible MEMS devices. The main soft technique employed is "soft" micromachining, which encompasses methods like soft lithography and other polymer processing techniques. Soft techniques offer advantages such as flexibility, ease of fabrication, and compatibility with biological systems. They have expanded the possibilities for fabricating MEMS devices with intricate structures and integrated functionalities [57]. Categorizing MEMS fabrication processes into hard and soft techniques helps differentiate between the more traditional and established methods and the newer, flexible approaches. Both categories contribute to the advancement of MEMS technology, enabling the creation of complex and innovative devices with a wide range of applications. These fabrication processes form the foundation for producing MEMS devices, allowing for the integration of mechanical, electrical, and optical components at a miniature scale. MEMS devices find applications in various fields, including aerospace, automotive, consumer electronics, healthcare, and telecommunications.

2.4 Thin-Film Deposition

Thin-film deposition encompasses various procedures for depositing thin films of various materials onto a surface. The properties of the deposited film depend highly on the deposition process, so selecting the appropriate technique is crucial when considering the desired application. Film specifications, such as internal stress, adhesion to the substrate, and conformity to the surface topography, are influenced by the deposition process. Certain process characteristics and criteria, such as contamination concerns or high temperatures, may limit the choice of substrates for deposition systems [58]. There are four primary methods of thin-film deposition: electrodeposition, physical vapor deposition, oxidation, and chemical vapor deposition [59]. Oxidation involves heating semiconductor substrates in the presence of H_2O vapor or O_2, resulting in the formation of high-quality thin oxide films, such as SiO_2 on silicon substrates. Chemical vapor deposition (CVD) encompasses all methods that utilize chemical reactions in a gas phase to deposit thin layers. Low-pressure CVD (LPCVD) and Plasma Enhanced CVD (PECVD) are two common techniques employed in microfabrication. Inorganic compounds like silicon dioxide, polycrystalline silicon (polysilicon), and silicon nitride are commonly deposited using these methods. Silicon dioxide may be thermally generated on a silicon wafer (thermal oxide) at high temperatures in an oxidation furnace or deposited by CVD [60].

Low-pressure chemical vapor deposition (LPCVD) is conducted within a pressure range of 0.01–1 Torr and temperatures between 560 and 900 °C. This process yields highly conformal films. In contrast, plasma-enhanced CVD (PECVD) takes place in plasma systems at lower temperatures but may lead to films with lower quality and conformity. Physical vapor deposition (PVD) methods utilize sputtering principles and evaporation. Evaporation involves heating the substance in a vacuum chamber, causing it to condense onto the substrate surface. Physical vapor deposition films are typically nonconformal due to the directionality of the deposition process. These methods are commonly used for depositing metals, although other materials can also be deposited. Electrodeposition, or electroplating, is a technique for creating thick metal structures by reducing metal ions in a solution onto a surface using an electrochemical process [59]. Different metals can be deposited using PVD techniques, such as gold, palladium, platinum, silver, and alloys. CVD can deposit various metals and metal compounds, including W, TiN, TiSi, WSi, and Cu. While metal CVD techniques are less common, they can improve step coverage or selectively deposit metals. Copper is often used as a connecting material due to its superior electromigration resistance and lower resistivity compared to aluminum, which has been the standard metallization choice in IC manufacturing for many years [61]. Polysilicon is commonly deposited using a low-pressure chemical vapor deposition (LPCVD) method with silane (SiH) as the gaseous precursor. Spin- or spray-coating are popular methods for depositing polymers like photoresists. Sensing applications may involve coating sensitive polymer films using similar processes. Dispensing or microcontact printing can also be utilized for depositing chemical or biological films [62].

2.4.1 Lithography

Lithography, derived from the Greek words for stone (lithos) and writing (gráphein), was introduced by Aloys Senefelder in 1796. Senefelder discovered that carved images on Bavarian limestone could be transferred to paper when properly inked and treated with chemicals [63–65]. Lithography is a key technology used in MEMS (MicroElectroMechanical Systems) for fabricating scaffolds [66]. It involves transferring three-dimensional patterns onto a flat surface, and it is a crucial step in the microfabrication process as it defines the shape of the device components [67, 68]. During lithography, a material is exposed to radiation and selectively removed in a developer solution after being locally modified by the radiation. There are three types of lithography based on the radiation source: Photolithography (optical), which uses UV light, particle-beam lithography, and X-beam Lithography, which utilizes particle or X-ray beams to modify the features of the material. When fabricating scaffolds, the main challenge lies in identifying radiation-sensitive materials that are also safe and biodegradable [63].

Lithography is a process commonly employed in various applications such as etching and doping. It involves transferring a computer-generated design onto a layer, typically glass or silicon, and subsequently etching an underlying thin film like oxide or nitride [69]. To create the desired geometric pattern, computer-aided design (CAD) is utilized, and a photomask is generated to guide the fabrication sequence. This comprises a photographic procedure resulting in a glass plate with the required design in the form of a thin chromium layer (about 100 nm). Once the photomask is ready, the lithography process can start [70–72]. The photomask, often made on a metallic thin film substrate (e.g., chromium, aluminum) opaque to UV light, is placed on a UV-transparent layer. The exposure method determines the resolution, and contact lithography, for example, involves direct contact between the photomask and the photoresist (PR). The photolithography method starts by spin-coating the layer with photosensitive resist materials, usually with a thickness of 2.5–0.5 μm, after the necessary materials have been deposited on it. After that, the coating is soft-baked to enhance adhesion and resistance and get rid of solvents. After that, the photoresist is exposed to UV light and the mask is aligned. The majority of MEMS fabrications typically use a g-line mercury source (436 nm) [73]. Comparable to the development of photographic film, the unexposed or exposed areas of the photoresist are washed away based on the kind of resist material (positive or negative). The desired pattern is then produced on the wafer after the resist is hard-baked to improve adherence. After removing the thin layer below, the photoresist is peeled using acetone or organic solvents. Because positive resist materials perform better in tiny geometries, they are often used in large-scale integration (VLSI) techniques (Fig. 2.1). The introduction of photolithography revolutionized the integrated circuit (IC) industry in the latter half of the twentieth century. Integrated circuits are now present in almost every electronic device we use today. As lithography processes advanced, transistors became thinner and smaller, resulting in faster and more efficient computers. Photolithography also played a significant role in the development of MicroElectroMechanical Systems (MEMS), which are now widely used in commercial products ranging from mechanical devices to biological applications, expanding the scope of IC technology [74–76]. To achieve smaller dimensions and higher aspect ratios, crucial for scaffold construction, various photolithography methods have been developed, including ion beams, electron beams, and X-rays [77]. These methods help in reducing the critical dimensions and improving the characteristics of the scaffolds.

2.5 Etching

2.5.1 Dry Etching

Etching methods play a crucial role in the fabrication of MEMS structures and integrated circuits (ICs). Etching is used to selectively remove layers or thin films after the lithography process, allowing the desired patterns to be realized. There are two main categories of etching methods: dry etching and wet etching [78–80]. Dry etching methods use a gas plasma as the etchant to remove materials [81]. They offer advantages such as anisotropy (directional etching), higher aspect ratio designs, and fine line patterning. Some common dry etching techniques include [82–84]:

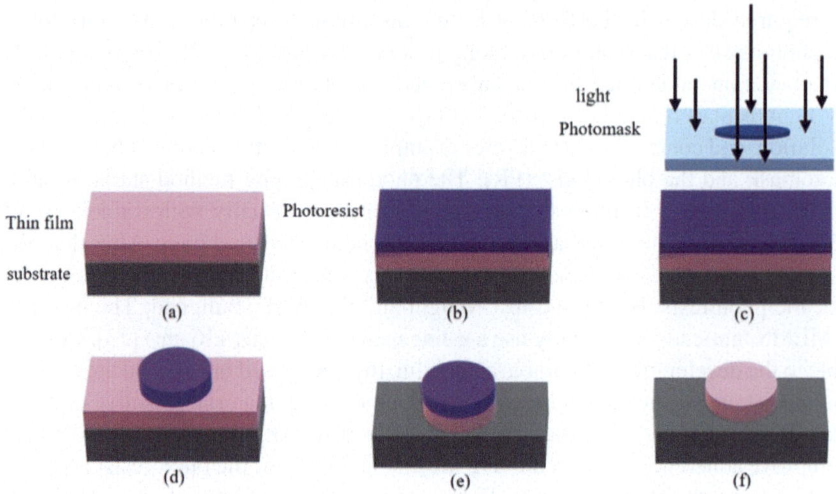

Fig. 2.1 Photolithographic stages with a positive photoresist are depicted schematically

1. Reactive Ion Etching (RIE): It involves a chemical reaction triggered by plasma to remove materials. Reactive ions generated in the plasma are directed towards the surface to be etched, resulting in directed etching properties.
2. Sputter Etching: It relies on a sputtering mechanism where the layer is bombarded by ions, causing material removal.
3. Vapor-Phase Etching: It involves the superficial dissolution of the target material when the wafer is placed in a chamber containing one or more gases that chemically interact with the material.
4. High-Pressure Plasma Etching: It produces highly reactive species that react with the target material, and the reaction products disperse, allowing further etching.
5. Ion Milling: It involves a balance between physical and chemical etching and requires adjustments to multiple parameters.

Deep Reactive Ion Etching (DRIE) is a specialized form of RIE that enables the creation of deep etchings with high aspect ratios. It employs a patented methodology developed by Bosch, which uses two distinct gas compositions to etch the layer and generate a sidewall passivation polymer. This passivation allows for etching aspect ratios of up to 50:1 [85].

2.5.2 *Wet Etching*

On the other hand, wet etching methods involve immersing the wafer in a fluid solution that dissolves the material. Wet etching is typically isotropic (non-directional),

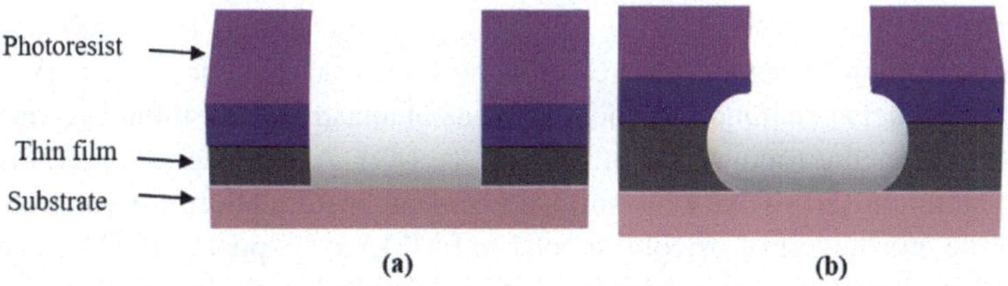

Fig. 2.2 Profile for **a** isotropic and **b** anisotropic etching using a photoresist mask

except for anisotropic wet etching of silicon, which can occur in an alkaline solution like potassium hydroxide. Some common wet etching techniques include [86, 87]:

1. Buffered Oxide Etch (BOE) or Diluted Hydrofluoric Acid (HF: H2O): It is used for isotropically etching silicon dioxide.
2. Anisotropic Wet Etching: It is employed to construct structures with a high aspect ratio. Common solutions for anisotropic wet etching include TMAH (Tetramethylammonium Hydroxide), KOH (Alkali-OH), and EDP (Ethylene Diamine Pyrochatechol).

TMAH selectively etches specific crystallographic planes of silicon, and the depth of the etched cavity is controlled by the concentration and temperature of the TMAH solution. Anisotropic wet etching is commonly used in the fabrication of microchambers, micronozzles, microchannels, and other structures. The choice of etching method, whether dry or wet, depends on the specific material and desired outcome. Dry etching methods offer anisotropy, higher aspect ratio designs, and fine line patterning, while wet etching methods provide options for high aspect ratio structures and selective etching. The dissolving rate in an isotropic etch is dependent on certain orientations and straight sidewalls, as illustrated in Fig. 2.2a, although alternative non-circular profiles may be achieved. In an anisotropic etch, the etchant attacks the matter at the same velocity in all directions, producing a semi-circular profile behind the mask, as shown in Fig. 2.2b. Both dry and wet etching techniques are essential in the creation of MEMS structures and integrated circuits, enabling the realization of intricate patterns and precise features for a wide range of applications.

2.5.3 *Direct Etching*

MEMS fabrication processes have advanced, allowing for the production of matrices with submicron or micron accuracy using silicon and polymeric materials. Zahn utilized linear plasma etching of a polymer, PMMA, to create high-resolution patterns for biomedical implants. In addition to plasma etching, physical etching methods can also be employed. Although slower and less selective, physical etching enables etching across a wider range of materials [88]. In physical etching, low-energy ions (5–0.010 keV) bombard the surface, leading to material ejection, substrate heating, and surface damage (referred to as ion etching or sputtering). This method, called "ion beam milling," uses a thick protective mask instead of a photoresist. Typically, etch rates fall between hundreds and thousands of angstroms per minute. It may also be used to alter the topography of the surface [89]. A secondary consequence of ion implantation (20–10 keV) is surface chemical modification. Researchers have studied the attachment of osteoblast cells to PMMA by exposing PMMA layers to Ar^+ or P^+ ion flows through masks. It was found that this technique caused nanoscale patterns to emerge on the surface of the PMMA, and that the surface modification improved cell adherence to the PMMA layer [90]. Additionally, the researchers proved that by choosing the right ions, such Ca or P, it was possible to modify the surface chemistry of biological materials. Specimens treated with Ar^+ had worse cell adhesion compared to PMMA specimens implanted with P^+ ions, for instance [91]. This method is characterised by low selectivity (almost 1:1 for the majority of materials) and slow etch rates (a few nanometers per minute) [7].

2.5.4 Substrate Bonding

Wafer bonding is widely used in microelectronics laboratories for manufacturing 3D sensors, SOI devices, and MEMS structures such as vacuum or gas cavities and channels. Two common applications of wafer bonding in MEMS are bonding the reference pressure chamber in pressure sensors and connecting components in accelerometers. In these cases, the bonding must be strong and airtight. Substrate (wafer) bonding, including glass-glass, silicon-glass, and silicon–silicon bonding, is a critical manufacturing method for microsystems. It is often used to build intricate three-dimensional structures that serve as functional parts and are integrated into the final microsystem packaging and encapsulation. Silicon-glass electrostatic, or anodic bonding, and silicon–silicon fusion bonding, often referred to as direct silicon bonding are two commonly utilized bonding techniques (Fig. 2.3). Thermal bonding requires high temperatures (around 1000 °C), while anodic bonding requires a strong electric field and relatively high temperatures (500 kV/m and 400 °C) [14]. Fusion bonding involves joining two cleaned silicon surfaces, which are then baked at high temperatures (800–300 °C) in nitrogen or oxygen to strengthen the bonds. The resulting composite has minimal thermal stress due to similar thermal expansion ratios of the two layers [13]. Fusion or direct silicon bonding is used for creating silicon-insulator (SOI) layers and micromechanical systems. This technique is typically employed to bond two silicon wafers that have not been coated with an oxide layer. The bonding process involves fully cleansing the silicon wafers coated with oxide or silicon, hydroxylating the surfaces in boiling nitric acid or HF to make them hydrophilic, and stacking the substrates starting from the center to prevent vacuum formation. The hydroxylated surfaces are held together by Van der Waals forces and hydrogen bonds, resulting in a robust link between the substrates. While it is possible to carry out these procedures at ambient temperature, a high-temperature annealing process (800–1200°C) is often required to improve the strength of the bonding.

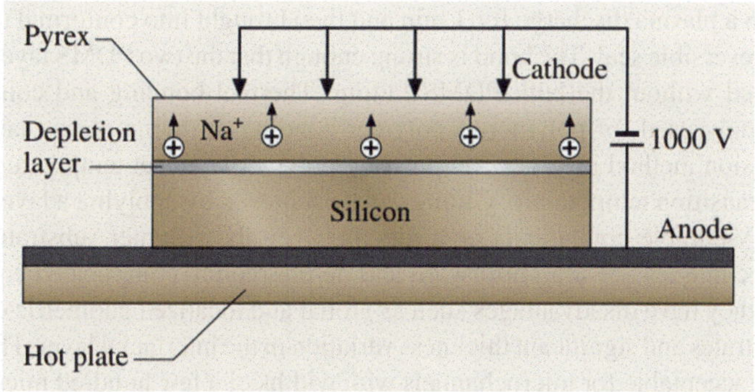

Fig. 2.3 Glass to silicon anodic bonding setup. Reprinted (adapted) with permission from [94]

Silicon fusion bonding has the benefit of being able to align the temperatures of the substrates precisely. Silicon-glass anodic bonding is widely used for bonding substrates in microsensor packaging and device production. The main benefit of this approach lies in its reduced bonding temperature. When a glass wafer, usually Pyrex 7740 because of its thermal expansion compatibility with silicon, is put atop a silicon wafer, the two are heated to a temperature range of 300–400 °C. Ultimately, the glass-silicon structure is exposed to an electrical potential of around 1000 V, where the glass is linked to the cathode. The bonding process starts with the application of voltage, creating an electrostatic force at the silicon-glass interface, which attracts and brings the substrates into contact. In addition to these techniques, other approaches using intermediary layers such as adhesives, glass frit, and eutectic bonding have been studied. The primary advantages of these techniques are their simplicity and the requirement of low-temperature annealing. However, a lengthy annealing process is a major disadvantage, often involving a 100-h vacuum storage period followed by at least 48 h of annealing to achieve a strong bond [92, 93].

For anodic bonding, materials having a thermal expansion coefficient comparable to silicon, such as Pyrex glass 7740, are used to reduce thermomechanical stress. To achieve void-free bonding, the wafer surfaces that are used for fusion and anodic bonding must be very smooth and clean. Surface roughness and quality become less important when using an average substrate for wafer bonding. Aside from adhesive layers and low melting temperature glasses, other possible intermediate bonding layers include solder films, metallic films such as gold (used in gold–silicon eutectic bonding), and other similar materials. Substrate bonding is often utilized to create structures that would be impossible or difficult to build otherwise (such as hermetically sealed large holes or enclosed channels) or to provide mechanical protection and support. Fusion and anodic bonding methods are not suitable for bonding polymeric materials, so alternative low-temperature layer bonding methods are required. Oxygen plasma bonding is a common technique for bonding PDMS layers in multilayer PDMS microfluidic systems. Figure 2.4 shows an example of enclosed microfluidic channels made from oxidized PDMS. The PDMS surfaces are oxidized in a plasma discharge for 1 min and then brought into conformal contact to form an irreversible seal. The bond is strong enough that the two PDMS layers cannot be separated without the bulk PDMS failing. Thermal bonding and compression are commonly used for polymer-to-polymer bonding of thermoplastic layers. The thermal fusion method involves compressing polymer layers at temperatures above the glass transition temperature. Gluing can be achieved by applying a layer of glue, such as UV-curable compounds or resins, between the polymer substrates. These bonding techniques are straightforward and do not require complicated equipment. However, they have disadvantages such as global and localized geometric distortion of the substrates and significant thickness variation in the interfacial layer. These limitations are acceptable for microchannels with widths of a few hundred micrometers. However, constructing micro- or nanosized channels becomes challenging due to the potential distortion of channel geometries. For bonding thermoplastic substrates, localized welding can be used instead. Infrared lasers, microwaves, or ultrasonics can be employed to melt the mating components adjacent to the microchannels. These techniques preserve the channel geometries and enable high-quality localized bonding by targeting specific areas or achieving uniform bonding across all mating surfaces. However, these efficient chip prototyping techniques require special designs of the mating parts and have not been widely used for microfluidic system fabrication [78].

2.6 Bulk Micromachining

Bulk micromachining is a well-established and commercially successful technique used in the manufacturing of Microelectromechanical Systems (MEMS). It involves selectively removing a layer of material, such as glass, GaAs, or silicon, to create micromechanical components like plates, membranes, and beams. This technique offers various advantages and has different variants [96]. There are two main methods used in silicon micromachining: wet etching and dry etching. Wet etching uses a chemical solution to selectively remove the material and is suitable for creating complex mechanical structures throughout the substrate's thickness. However, achieving high precision in wet etching requires careful control of the etching conditions and mask layout [97]. Dry etching, on the other hand, involves using plasma or reactive gases to remove the material. It offers the advantage of producing high-quality structures with precise dimensions and upright sidewalls. Dry etching techniques are compatible with active microelectronics and provide more flexibility in process integration [98].

In wet etching, the final shape of the structures is determined by the initial mask design, the etching method, and the wafer's properties. Wet anisotropic etching can create structures like lenses, fibers, and V-grooves. It requires precise control of the etching conditions and mask layout. Wet etching can also be used to create cantilevers and other structures with upright sidewalls. The wet anisotropic etching can produce shapes like rhombic channels, cantilevers, and V-grooves [99]. Dry etch bulk silicon

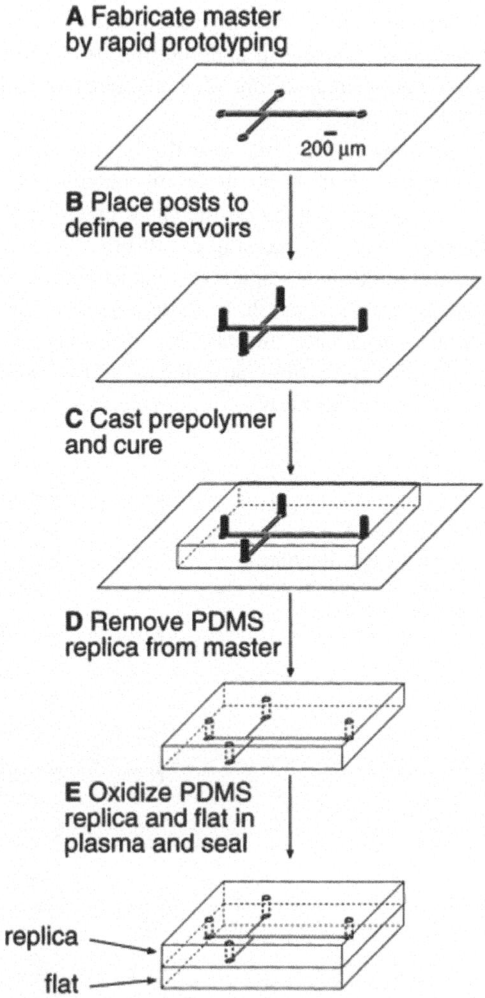

Fig. 2.4 The fabrication of an enclosing microfluidic channel in oxidized PDMS is described. **a** In photolithography, high-resolution transparency containing the channel design was utilized as the mask to generate a positive photoresist relief on a silicon wafer. **b** Reservoirs were defined on the wafer using glass posts. **c** After that, PDMS was cast onto the silicon wafer and cured for 1 h at 65 °C. **d** The glass posts were removed, and the polymer duplicate of the master with a negative relief of channels was peeled away from the silicon wafer. **e** This was done to oxidize the replica and the flat slab of PDMS for 1 min using a plasma discharge. Between the PDMS layers, an irreversible seal was created. Reprinted (adapted) with permission from [95]. Copyright © 1998, American Chemical Society [95]

micromachining techniques provide an alternative to wet techniques. Developed in the mid-1990s, these techniques utilize anisotropic dry etching processes and offer high-feature proportion upright structures. Dry etching involves using plasma or reactive gases to remove the material and offers the advantage of producing high-quality structures with precise dimensions and upright sidewalls. Dry etching methods, such as XeF2 vapor phase etch and anisotropic/anisotropic etch combinations, can create upright structures. Dry etching techniques have less compatibility concerns with active microelectronics and can be used in conjunction with on-chip electronics or in isolation [100]. Figure 2.5 demonstrates the process of the Single-Crystal Reactive Etching and Metallization (SCREAM) method. It starts with an anisotropic silicon etch with an oxide mask, followed by oxide deposition and subsequent etching to remove the oxide from the trench bottoms. Anisotropic silicon etching finalizes the structures, resulting in undercutting. If electrostatic incitement is required, a metal can be deposited on the sidewalls, bottom, and top of the cavity created below the microstructure [94].

Bulk micromachining can also be combined with microelectronics manufacturing techniques to integrate micromechanical elements with mixed circuits (Bipolar, BiCMOS, or CMOS) on the same layer. This integration enables improved interface and signal processing capabilities. However, manufacturing concerns such as process yield and compatibility need to be carefully considered. Overall, bulk micromachining is a versatile technique used in MEMS manufacturing, offering various methods to create intricate micromechanical components. Wet and dry etching techniques have their advantages and considerations, and their integration with microelectronics allows for the creation of more advanced devices [101].

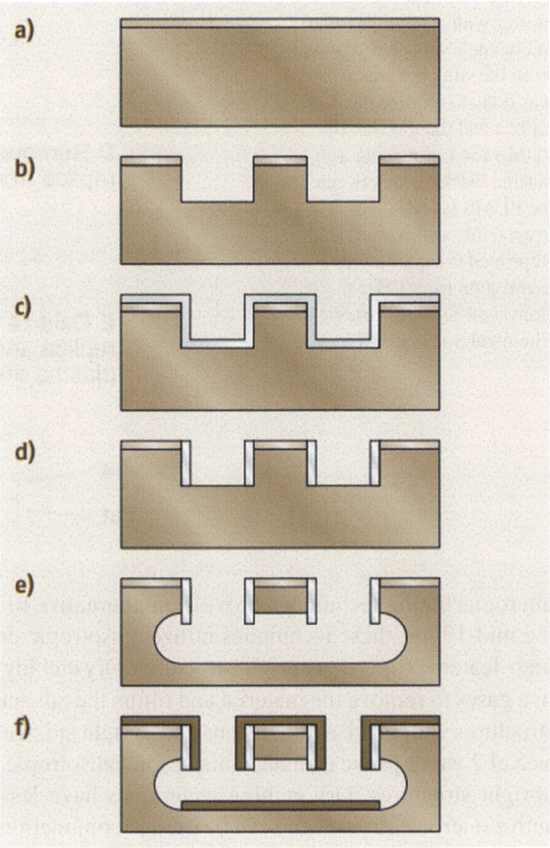

Fig. 2.5 Shows a cross-part of the SCREAM method. The stages are described in detail in the text. Reprinted (adapted) with permission from Fig. 26 [94]

2.7 Surface Micromachining

Surface micromachining is a crucial process in MEMS microfabrication, allowing for the creation of movable microstructures on top of a silicon layer. This technique involves etching thin films, precise patterning, and deposition, resulting in structures typically ranging from 1 to 100 μm in thickness. The key advantage of surface micromachining is its ability to achieve incredibly small sizes. Additionally, combining micromachined components with on-chip circuits to enhance functionality is a straightforward process. However, the nonlinearity of the surface limits the maximum number of layers that can be deposited [102]. The fundamental procedure of surface micromachining involves producing and patterning a silicon layer, followed by the formation and patterning of a sacrificial layer (as shown in Fig. 2.6). Polysilicon became the preferred material in the 1980s and remains widely used today, although other materials like silicon nitride, aluminum, and silicon dioxide are also significant in surface micromachining. The choice of the sacrificial layer depends on the mechanical layer used, ensuring that the victim layer can be etched without significantly affecting the mechanical substrate or layer. A commonly employed combination is a mechanical layer of silicon nitride or polysilicon with an oxide victim layer [94].

Stiction, which can occur during the manufacturing process or throughout the device's lifespan, is an important factor in the processing and design of micromachined surface devices. Methods for reducing stiction include using lower interface tension fluids, employing steam or dry phase etchants, implementing geometrical modifications, applying organic slim-film surface modifications or self-assembled monolayers (SAM), and utilizing freeze-drying. In-use stiction during the device's

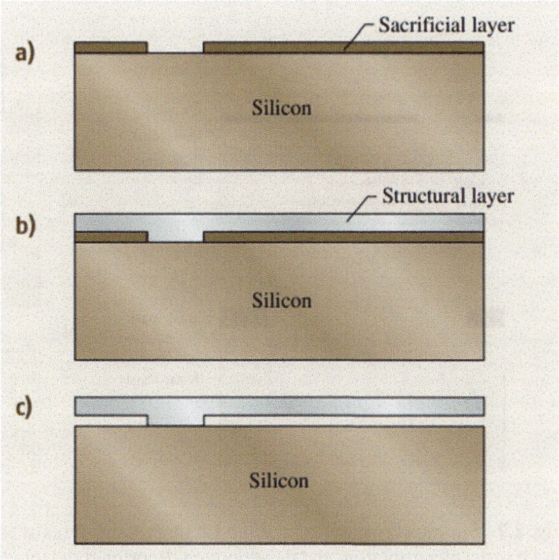

Fig. 2.6 Micromachining process basic surface manufacturing. Reprinted (adapted) with permission from Fig. 31 [94]

operational lifespan can be caused by direct chemical bonding, moisture condensation, or electrostatic charge buildup. Surface passivation using self-assembled organic slim layers or monolayers can minimize surface energy, capillary forces, and electrostatic forces, especially when directly applied to the semiconductor without an interfering oxide substrate [103].

2.8 High-Aspect-Ratio Micromachining

High-aspect-ratio micromachining is a specialized manufacturing technique used in MEMS (Micro-electromechanical Systems) to create microdevices with intricate features that would be challenging to achieve using traditional 2D surface or bulk technologies. This process focuses on developing devices with high actuation or sensing performance [104]. The most commonly employed method for high-aspect-ratio micromachining is called LIGA, which stands for Lithographie, Abformung, Galvanoformung (German for Lithography, Molding, Electroplating). LIGA was initially developed in Germany and has since been refined to make it more cost-effective [105]. The LIGA process involves several fundamental manufacturing steps. First, X-ray lithography is performed on a thick, resistant substrate. This lithography step defines the desired patterns on the substrate. Next, electroplating is used to fill the exposed regions with metal, resulting in the formation of three-dimensional structures. Finally, the resist material is removed, leaving behind the metal structures with vertical sidewalls (Fig. 2.7) [94].

One of the key advantages of LIGA is its ability to achieve high aspect ratios, which refer to the ratio of the structure's height to its lateral dimensions. With LIGA,

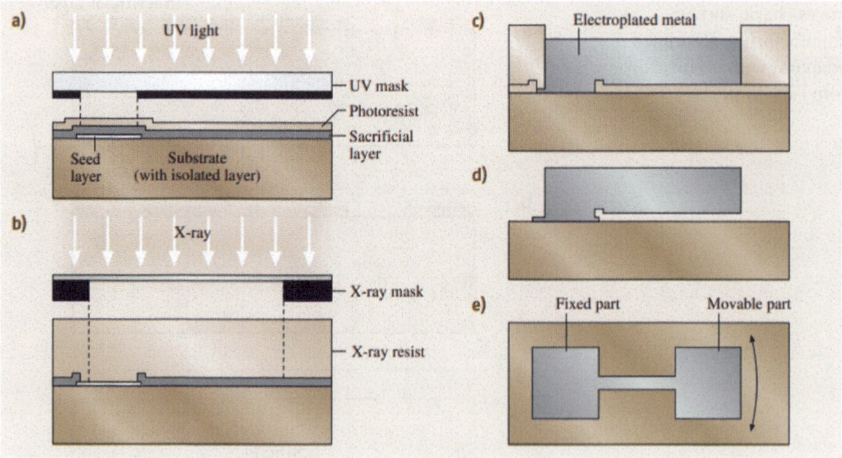

Fig. 2.7 Shows a cross-section of the LIGA process. **c** resist stripping, **a** resist X-beam exposure, and **b** resist electroplating and development. Reprinted (adapted) with permission from Fig. 38 [94]

aspect ratios greater than 20:1 and even exceeding 100:1 can be achieved. This means that the structures can have significantly greater height compared to their lateral dimensions.

To accomplish such high aspect ratios, LIGA utilizes high-energy X-rays generated by synchrotron radiation. The short wavelengths of these X-rays allow them to penetrate thick substrates without significant dispersion, enabling the creation of tall and narrow structures.

While LIGA has been widely used in research settings, its adoption in industrial manufacturing has been limited due to the high cost and restricted availability of synchrotron X-ray equipment. As a more cost-effective alternative, a technique known as "UV-LIGA" or "poor man's LIGA" has been developed. UV-LIGA employs standard contact lithography equipment and negative photoresist materials, offering a more accessible approach to achieving relatively high aspect ratios, typically exceeding 20:1.

In summary, high-aspect-ratio micromachining, particularly through methods like LIGA, enables the fabrication of microdevices with intricate three-dimensional structures. By utilizing X-ray lithography and electroplating, these techniques achieve high aspect ratios, allowing for the creation of microstructures with greater height compared to their lateral dimensions. While LIGA is the most widely used method, UV-LIGA provides a more cost-effective alternative for achieving relatively high aspect ratios.

2.9 Soft Micromachining

As previously mentioned, soft materials such as gels and polymers offer distinct advantages over glass and silicon in the field of biomedical MEMS. Soft lithography, which involves creating microstructures and micropatterns using soft materials, relies on two ancient methods: embossing and molding [106]. Silicone rubber, or polydimethylsiloxane (PDMS), is the material of choice for micro-molding. PDMS is perfect for microfluidics because it is inexpensive, biocompatible, and has great sealing properties [107–110]. It also allows for easy layering, enabling the creation of complex structures. Additionally, the elastic properties of thin PDMS diaphragms allow for significantly larger deflections compared to rigid materials. Figure 2.8 illustrates the fabrication process of PDMS structures using micromolding [106].

The procedure begins with the creation of a mold using one of the hard micromachining processes outlined earlier [111–115]. Depending on the material of the mold and its topography, surface treatment is usually suggested to make demolding easier. Once the micro mold is prepared, a prepolymer solution (usually a combination of a crosslinking agent and PDMS linear polymer) is poured onto it. This procedure may need to be done under a vacuum in certain circumstances to remove air bubbles from molds with high aspect ratios. Following the completion of polymerization, which may vary from one to twenty-four hours according to the curing temperature, the PDMS substrate is carefully removed. Costs may be greatly decreased by

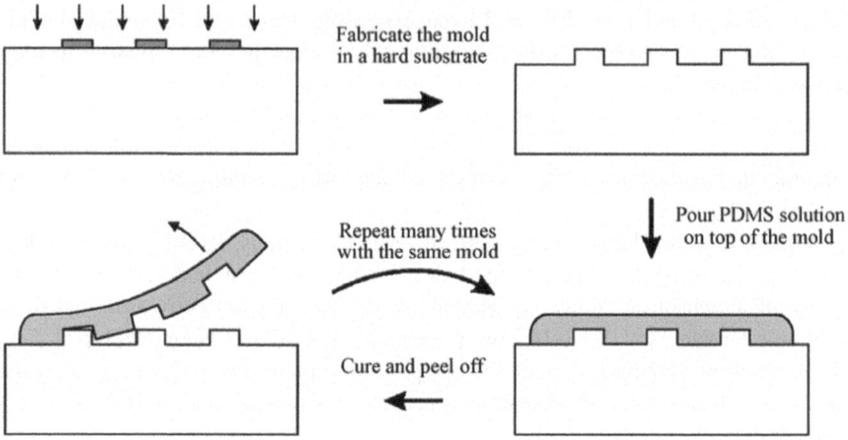

Fig. 2.8 Micromolding process [106]

using the mold repeatedly. In addition to PDMS, soft lithography has been utilized to create monolithic PDMS blocks with various configurations of microchannels by bonding multiple patterned PDMS substrates together. By interposing a slender PDMS diaphragm amongst two microchannels, it is possible to create peristaltic pumps and pneumatically actuated valves by exerting pressure on the upper channel. As a result, the diaphragm deflects, thereby obstructing the flow of liquid in the lower channel. The extensibility of the PDMS diaphragm is determined by variables such as the concentration of crosslinker in the prepolymer, the thickness of the diaphragm, and the aspect ratio. The extensibility of the PDMS diaphragm is influenced by factors such as the amount of crosslinker in the prepolymer, thickness, and aspect ratio. Moreover, PDMS exhibits excellent adhesion to silicon nitride and glass, and surface preparation with HCl or plasma facilitates bonding with other materials [116].

An alternate method to microfabrication using glass and silicon that is less expensive is the warm embossing of polycarbonate and PMMA. For instance, a PMMA microfluidic chip for DNA detection and separation was created using quartz templates with a negative image of microchannels. The process is schematically depicted in Fig. 2.9. Initially, the master or stamp is brought into contact with the plastic, and then the layer is heated to reach its glass transition temperature. An evenly distributed pressure is applied, and the layer is allowed to cool before removing the stamp. Warm embossing has also demonstrated success in nanoscale manufacturing, particularly in the field of optical technology. It has been employed to produce optical waveguides, microlenses, diffraction gratings, and other optical components [106, 117].

Beebe et al. have introduced "microfluidic tectonics" (AFT), a collection of methods for micromachining polymeric materials. This method, particularly advantageous for rapid prototyping of MEMS structures, relies on photopolymerization with

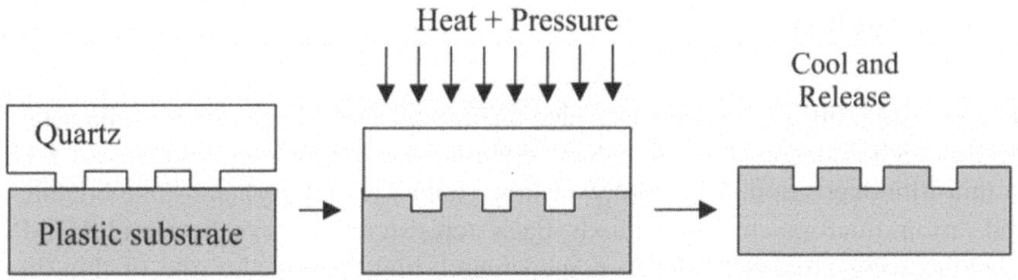

Fig. 2.9 Showing the hot embossing technique. Steps are described in the text [106]

a series of masks that prevent polymerization in specific areas. Following a complete photopolymerization stage, unreacted monomers with retained polymerization characteristics can be washed away, depending on whether the polymer is glassy or rubbery. Multiple features can be created simultaneously through a single mask polymerization or sequentially through photopolymerization sequences using different masks. The combination of the crosslinker tetramethylene glycol dimethacrylate and the monomers isobornyl acrylate is well-suited for rapid polymerization under UV light in the presence of the photoinitiator Irgacure 651, as they polymerize quickly with minimal volume change. Photopolymerization of soft materials like hydrogels is also possible when using a mask [118].

After UV exposure and rinsing, a mask placed directly over the prepolymer solution will result in a microfluidic channel. The channel is then filled with an aqueous hydrogel solution and photopolymerized under a grating that periodically blocks UV light. After rinsing, the outcome is a tube with void spaces separated by hydrogel at regular intervals. Beebe and Eddington discuss more complex hydrogel-AFT devices in their work. Microfluidic tectonics offers several advantages in terms of simplicity and processing flexibility. However, there are limitations to AFT. The reflection, diffraction, and "penumbra" of polymerization around the directly irradiated area, as well as the scattering of incoming UV light, restrict the packing density and size of structures. AFT structures under 50 nm are notoriously challenging to manufacture. This constraint may be addressed by switching to more costly multiphoton sources that provide nanoscale resolution or by forgoing device depth in favor of increased lateral clarity. Another constraint of AFT is the geometric repeatability of structures produced using soft stimuli-sensitive materials like hydrogels. Since these structures expand after polymerization when exposed to different chemical conditions, their final shape is influenced by a wider range of variables compared to solid-phase lithography. While AFT has primarily been used for rapid prototyping, its compatibility with other techniques and the versatility of materials makes it an attractive platform for future research [119].

2.10 Conclusion

In conclusion, this chapter has provided a comprehensive exploration of microfabrication techniques used in MEMS production. It covers essential processes such as thin-film deposition, lithography, etching (both wet and dry), substrate bonding, and various micromachining methods. Each step's significance in the overall MEMS manufacturing process has been demonstrated, highlighting the role of thin-film deposition in creating functional layers, the importance of lithography in defining structures, the selective material removal capabilities of etching, the integration possibilities through substrate bonding, and the versatility of different micromachining techniques. Furthermore, the chapter has highlighted the emergence of soft micromachining, which has opened up new possibilities for flexible and deformable MEMS structures. This advancement expands the potential applications of MEMS technology, particularly in areas such as biomedicine and drug delivery systems. The comprehensive understanding gained from this chapter will contribute to the advancement of MEMS technology, enabling the design and fabrication of innovative devices with enhanced performance and expanded applications. By mastering the microfabrication techniques discussed, researchers and engineers can push the boundaries of miniaturization and unlock the full potential of MEMS in various industries.

References

1. Chircov, C., Grumezescu, A.M.: Microelectromechanical systems (MEMS) for biomedical applications. Micromachines **13**, 164 (2022). https://doi.org/10.3390/mi13020164
2. Mishra, M.K., et al.: MEMS technology: a review. J. Eng. Res. Rep. **4**(1), 1–24 (2019)
3. Algamili, A.S., et al.: A review of actuation and sensing mechanisms in MEMS-based sensor devices. Nanoscale Res. Lett. **16**(1), 16 (2021)
4. Hassanin, H., et al.: Microadditive manufacturing technologies of 3D microelectromechanical systems. Adv. Eng. Mater. **23**(12), 2100422 (2021)
5. De Pasquale, G.: Additive manufacturing of micro-electro-mechanical systems (MEMS). Micromachines **12**, 1374 (2021). https://doi.org/10.3390/mi12111374
6. Hajare, R., Reddy, V., Srikanth, R.: MEMS based sensors—a comprehensive review of commonly used fabrication techniques. Mater. Today Proc. **49**, 720–730 (2022)
7. Feynman, R.: There's plenty of room at the bottom. In Feynman and computation (pp. 63–76). CRC Press (2018).
8. Petersen, K.E.: Silicon as a mechanical material. Proceedings of the IEEE **70**(5), 420–457 (1982). https://doi.org/10.1109/PROC.1982.12331
9. Silverio, V., de Freitas, S.C.: Microfabrication techniques for microfluidic devices. In: Galindo-Rosales, F.J. (ed.) Complex Fluid-Flows in Microfluidics, pp. 25–51. Springer International Publishing: Cham (2018)
10. Pillai, G., Li, S.S.: Piezoelectric MEMS resonators: a review. IEEE Sens. J. **21**(11), 12589–12605 (2021)
11. Cao, T., Hu, T., Zhao, Y.: Research status and development trend of MEMS switches: a review. Micromachines **11**, 694 (2020). https://doi.org/10.3390/mi11070694
12. Basu, A.K., et al.: MEMS Applications in Biology and Healthcare. AIP Publishing LLC, Melville (2021)

13. Ren, Z., et al.: Leveraging of MEMS technologies for optical metamaterials applications. Adv. Opt. Mater. **8**(3), 1900653 (2020)
14. Chircov, C., Grumezescu, A.M.: Microelectromechanical systems (MEMS) for biomedical applications. Micromachines (Basel) **13**(2), 164 (2022)
15. Hossain, N., et al.: Advances of materials science in MEMS applications: a review. Res. Eng. **22**, 102115 (2024)
16. Obaid, S.N., Chen, Z., Lu, L.: Advanced electrical and optical microsystems for biointerfacing. Adv. Intell. Syst. **2**(9), 2000091 (2020)
17. Sophia, N., Oh Seok, K.: Micro-electromechanical systems-based sensors and their applications. Appl. Sci. Conver. Technol. **31**(2), 40–45 (2022)
18. Kumar, A., et al.: Recent development and futuristic applications of MEMS based piezoelectric microphones. Sens. Actuators A **347**, 113887 (2022)
19. Lee, Y.-C., et al.: CMOS-MEMS technologies for the applications of environment sensors and environment sensing hubs. J. Micromech. Microeng. **31**(7), 074004 (2021)
20. Zhou, G., et al.: MEMS gratings and their applications. Int. J. Optomechatronics **15**(1), 61–86 (2021)
21. Hartzell, A.L., Da Silva, M.G., Shea, H.R.: MEMS Reliability. Springer Science & Business Media (2010)
22. Khan, M.S., et al.: MEMS sensors for diagnostics and treatment in the fight against COVID-19 and other pandemics. IEEE Access **9**, 61123–61149 (2021)
23. Zhu, J., et al.: Development trends and perspectives of future sensors and MEMS/NEMS. Micromachines **11**, 7 (2020). https://doi.org/10.3390/mi11010007
24. Garcia, L., et al.: The role of soft robotic micromachines in the future of medical devices and personalized medicine. Micromachines **13**, 28 (2022). https://doi.org/10.3390/mi13010028
25. Salahandish, R., et al.: Label-free ultrasensitive detection of breast cancer miRNA-21 biomarker employing electrochemical nano-genosensor based on sandwiched AgNPs in PANI and N-doped graphene. Biosens. Bioelectron. **120**, 129–136 (2018)
26. Salahandish, R., et al.: Nano-biosensor for highly sensitive detection of HER2 positive breast cancer. Biosens. Bioelectron. **117**, 104–111 (2018)
27. Shahbazi, N., Zare-Dorabei, R., Naghib, S.M.: Design of a ratiometric plasmonic biosensor for herceptin detection in HER2-positive breast cancer. ACS Biomater. Sci. Eng. **8**(2), 871–879 (2022)
28. Salahandish, R., et al.: Sandwich-structured nanoparticles-grafted functionalized graphene based 3D nanocomposites for high-performance biosensors to detect ascorbic acid biomolecule. Sci. Rep. **9**(1), 1226 (2019)
29. Tabeling, P.: Introduction to Microfluidics. Oxford University Press (2023)
30. Naghib, S.M., et al.: A highly sensitive biosensor based on methacrylated graphene oxide-grafted polyaniline for ascorbic acid determination. Nanotechnol. Rev. **9**(1), 760–767 (2020)
31. Sadeghi, M., et al.: A highly sensitive nanobiosensor based on aptamer-conjugated graphene-decorated rhodium nanoparticles for detection of HER2-positive circulating tumor cells. Nanotechnol. Rev. **11**(1), 793–810 (2022)
32. Mohammadpour, Z., Naghib, S.M.: Chapter 15—smart nanosensors for intelligent packaging. In: Thomas, S., et al. (eds.) Nanosensors for Smart Manufacturing, pp. 323–346. Elsevier (2021)
33. Salahandish, R., et al.: A novel graphene-grafted gold nanoparticles composite for highly sensitive electrochemical biosensing. IEEE Sens. J. **18**(6), 2513–2519 (2018)
34. Tian, W.-C., Finehout, E.: Introduction to microfluidics. In: Microfluidics for Biological Applications, pp. 1–34. Springer (2008)
35. Cottet, J., Renaud, P.: Chapter 1—introduction to microfluidics. In: Chappel, E. (ed.) Drug Delivery Devices and Therapeutic Systems, pp. 3–17. Academic Press (2021)
36. Dietzel, A.: A brief introduction to microfluidics. In: Dietzel, A. (ed.) Microsystems for Pharmatechnology: Manipulation of Fluids, Particles, Droplets, and Cells, pp. 1–21. Springer International Publishing, Cham (2016)

37. Verma, N., et al.: Chapter one—an introduction to microfluidics and their applications. In: Pandya, A., Singh, V. (eds.) Progress in Molecular Biology and Translational Science, pp. 1–14. Academic Press (2022)
38. Ottino, J.M., et al.: Introduction: mixing in microfluidics. Philos. Trans. R. Soc. Lond. Ser. A Math. Phys. Eng. Sci. **362**(1818), 923–935 (2004)
39. Chen, K., Hugh Fan, Z.: Introduction to Microfluidics, in Circulating Tumor Cells, pp. 33–50 (2016)
40. Paegel, B.M., Blazej, R.G., Mathies, R.A.: Microfluidic devices for DNA sequencing: sample preparation and electrophoretic analysis. Curr. Opin. Biotechnol. **14**(1), 42–50 (2003)
41. Zhang, C., Xing, D.: Single-molecule DNA amplification and analysis using microfluidics. Chem. Rev. **110**(8), 4910–4947 (2010)
42. Snider, A., et al.: A microfluidics workflow for sample preparation for next-generation DNA sequencing. SLAS Technol. Transl. Life Sci. Innov. **24**(2), 196–208 (2018)
43. Grayson, A.C.R., et al.: A BioMEMS review: MEMS technology for physiologically integrated devices. Proc. IEEE **92**(1), 6–21 (2004)
44. Amirouche, F., Zhou, Y., Johnson, T.: Current micropump technologies and their biomedical applications. Microsyst. Technol. **15**(5), 647–666 (2009)
45. Bogue, R.: Recent developments in MEMS sensors: a review of applications, markets and technologies. Sens. Rev. **33**(4), 300–304 (2013)
46. Bishop, D.J., Giles, C.R., Austin, G.P.: The Lucent LambdaRouter: MEMS technology of the future here today. IEEE Commun. Mag. **40**(3), 75–79 (2002)
47. Lee, K.B.: Principles of Microelectromechanical Systems. Wiley (2011)
48. Tsao, C.-W.: Polymer microfluidics: simple, low-cost fabrication process bridging academic lab research to commercialized production. Micromachines **7**, 225 (2016). https://doi.org/10.3390/mi7120225
49. Yeo, L.Y., et al.: Microfluidic devices for bioapplications. Small **7**(1), 12–48 (2011)
50. Matellan, C., del Río Hernández, A.E.: Cost-effective rapid prototyping and assembly of poly(methyl methacrylate) microfluidic devices. Sci. Rep. **8**(1), 6971 (2018)
51. Gale, B.K., et al.: A review of current methods in microfluidic device fabrication and future commercialization prospects. Inventions **3**, 60 (2018). https://doi.org/10.3390/inventions3030060
52. Madou, M.J.: Fundamentals of Microfabrication and Nanotechnology, Three-Volume Set. CRC Press (2018)
53. Franssila, S.: Introduction to Microfabrication. Wiley (2010)
54. Huff, M.: Process Variations in Microsystems Manufacturing. Springer (2020)
55. Mahmud, M.A.P., et al.: Advances in MEMS and microfluidics-based energy harvesting technologies. Adv. Mater. Technol. **7**(7), 2101347 (2022)
56. Fitzgerald, A.M., White, C.D., Chung, C.C.: MEMS product development, vol. 10, pp. 973–978. Springer International Publishing, Cham (2021)
57. Kumar, S., et al.: Additive manufacturing as an emerging technology for fabrication of microelectromechanical systems (MEMS). J. Micromanufact. **2**(2), 175–197 (2019)
58. Seshan, K., Schepis, D.: Handbook of Thin Film Deposition. William Andrew (2018)
59. Morosanu, C.E.: Thin Films by Chemical Vapour Deposition, vol. 7. Elsevier (2016)
60. Sun, L., et al.: Chemical vapour deposition. Nat. Rev. Methods Primers **1**(1), 5 (2021)
61. Franz, G.: Plasma enhanced chemical vapor deposition of organic polymers. Processes **9**, 980 (2021). https://doi.org/10.3390/pr9060980
62. Boden, S.A., Temple, T.L.: Light capture. In: Irvine, S.J.C. (ed.) Materials Challenges, p. 0. The Royal Society of Chemistry (2014)
63. Ryu, Y.K., Rodrigo, J.M.: Scanning Probe Lithography: Fundamentals, Materials, and Applications. CRC Press (2022)
64. Sievers, G.: Learning how to print in Colonial North India: the Nizami Press in Budaun and the First Urdu manual on the art of lithography. Philol. Encount. **8**(1), 73–109 (2023)
65. Scardamaglia, A.: A legal history of lithography. Griffith Law Rev. **26**(1), 1–27 (2017)

66. Nadine, S., et al.: Advances in microfabrication technologies in tissue engineering and regenerative medicine. Artif. Organs **46**(7), E211–E243 (2022)
67. Puryear, J.R. III, Yoon, J.-K., Kim, Y.: Advanced fabrication techniques of microengineered physiological systems. Micromachines **11**, 730 (2020). https://doi.org/10.3390/mi11080730
68. Blachowicz, T., Ehrmann, A.: 3D Printed MEMS technology—recent developments and applications. Micromachines **11**, 434 (2020). https://doi.org/10.3390/mi11040434
69. Middya, S., et al.: Chapter 8—Lithography and electrodes. In: Das, S., Dhara, S. (eds.) Chemical Solution Synthesis for Materials Design and Thin Film Device Applications, pp. 277–307. Elsevier (2021)
70. Chen, B., Feng, B., Lou, J.J.-C.: Mask manufacturing equipment. In: Wang, Y., et al. (eds.) Handbook of Integrated Circuit Industry, pp. 1285–1325. Springer Nature Singapore, Singapore (2024)
71. Sturtevant, J., Capodieci, L.: Design for manufacturing and design process technology co-optimization. In: Microlithography, pp. 293–326. CRC Press (2020)
72. Hines, D.R., et al.: MEMS lithography and micromachining techniques. In: MEMS Materials and Processes Handbook, pp. 667–753 (2011)
73. Hirai, Y., et al.: A three-dimensional microstructuring technique exploiting the positive photoresist property. J. Micromech. Microeng. **20**(6), 065005 (2010)
74. Dwivedi, S.: Fabrication techniques and materials for bio-MEMS. In: Guha, K., et al. (eds.) MEMS and Microfluidics in Healthcare: Devices and Applications Perspectives, pp. 101–141. Springer Nature Singapore, Singapore (2023)
75. Lall, J., Zappe, H.: MEMS-compatible structuring of liquid crystal network actuators using maskless photolithography. Smart Mater. Struct. **31**(11), 115014 (2022)
76. Wang, J., et al.: Preparation of microneedle array mold based on MEMS lithography technology. Micromachines **12**, 1–23 (2021). https://doi.org/10.3390/mi12010023
77. Handrea-Dragan, I.M., et al.: Patterning at the micro/nano-scale: Polymeric scaffolds for medical diagnostic and cell-surface interaction applications. Colloids Surf. B **218**, 112730 (2022)
78. He, Y., et al.: Micro electro-mechanical systems (MEMS). In: Wang, Y., et al. (eds.) Handbook of Integrated Circuit Industry, pp. 895–911. Springer Nature Singapore, Singapore (2024)
79. Guo, H., et al.: A review on the mainstream through-silicon via etching methods. Mater. Sci. Semicond. Process. **137**, 106182 (2022)
80. Kurmendra, Kumar, R.: Materials selection approaches and fabrication methods in RF MEMS switches. J. Electron. Mater. **50**(6), 3149–3168 (2021)
81. Lim, N., et al.: Dry etching performance and gas-phase parameters of $C_6F_{12}O + Ar$ plasma in comparison with $CF_4 + Ar$. Materials **14**, 1595 (2021). https://doi.org/10.3390/ma14071595
82. Lotito, V., Zambelli, T.: Playing with sizes and shapes of colloidal particles via dry etching methods. Adv. Coll. Interface. Sci. **299**, 102538 (2022)
83. Gasvoda, R.J., et al.: Etch selectivity during plasma-assisted etching of SiO_2 and SiNx: transitioning from reactive ion etching to atomic layer etching. J. Vac. Sci. Technol. A **38**(5), 050803 (2020)
84. Islam, R., Cui, B., Miao, G.-X.: Dry etching strategy of spin-transfer-torque magnetic random access memory: a review. J. Vac. Sci. Technol. B **38**(5), 050801 (2020)
85. Dirdal, C.A., et al.: Towards high-throughput large-area metalens fabrication using UV-nanoimprint lithography and Bosch deep reactive ion etching. Opt. Express **28**(10), 15542–15561 (2020)
86. Gablech, I., et al.: Infinite selectivity of wet SiO_2 etching in respect to Al. Micromachines **11** (2020). https://doi.org/10.3390/mi11040365
87. Lee, Y.-C., et al.: Fabrication and function examination of PZT-based MEMS accelerometers. Ceram. Int. **47**(17), 24458–24465 (2021)
88. Zahn, J. D., et al.: A direct plasma etch approach to high aspect ratio polymer micromachining with applications in bioMEMS and CMOS-MEMS. In Technical Digest. MEMS 2002 IEEE International Conference. Fifteenth IEEE International Conference on Micro Electro Mechanical Systems (Cat. No. 02CH37266) (pp. 137–140). IEEE (2002, January)

89. Madou, M. J.: Fundamentals of microfabrication: the science of miniaturization. CRC press (2018)
90. He, W., et al.: Micro/nanomachining of polymer surface for promoting osteoblast cell adhesion. Biomed. Microdev., **5**, 101–108 (2003)
91. Dorsey, K.J.: Atomic Layer Deposition for Mechanical, Magnetic, and Robotic Systems. Cornell University (2020)
92. Xu, J., et al.: Progress in wafer bonding technology towards MEMS, high-power electronics, optoelectronics, and optofluidics. Int. J. Optomechatronics **14**(1), 94–118 (2020)
93. Xu, M., et al.: Design and fabrication of an absolute pressure MEMS capacitance vacuum sensor based on silicon bonding technology. Vacuum **186**, 110065 (2021)
94. Ziaie, B., A. Baldi, and M.Z. Atashbar, Introduction to micro-/nanofabrication. In: Springer Handbook of Nanotechnology, pp. 231–269 (2010)
95. Duffy, D.C., et al.: Rapid prototyping of microfluidic systems in poly(dimethylsiloxane). Anal. Chem. **70**(23), 4974–4984 (1998)
96. Pal, P., et al.: High speed silicon wet anisotropic etching for applications in bulk micromachining: a review. Micro Nano Syst. Lett. **9**(1), 4 (2021)
97. Mareev, E., et al.: Single-shot femtosecond bulk micromachining of silicon with mid-IR tightly focused beams. Sci. Rep. **12**(1), 7517 (2022)
98. Li, Y., et al.: A high-sensitivity resonant differential pressure microsensor based on bulk micromachining. IEEE Sens. J. **21**(7), 8927–8934 (2021)
99. Wang, L., et al.: Two-step femtosecond laser etching for bulk micromachining of 4H-SiC membrane applied in pressure sensing. Ceram. Int. **48**(9), 12359–12367 (2022)
100. Zhan, S., et al.: A novel approach for bulk micromachining of 4H-SiC by tool-based electrolytic plasma etching in HF-free aqueous solution. J. Eur. Ceram. Soc. **41**(10), 5075–5087 (2021)
101. Li, X., Yang, H.: Bulk micromachining. In: 3D and Circuit Integration of MEMS, pp. 13–48 (2021)
102. Singh, R.K., et al.: Introduction to MEMS and microfluidics. In: Singh, R.K., et al. (eds.) Advances in MEMS and Microfluidic Systems, pp. 1–10. IGI Global, Hershey, PA, USA (2023)
103. Madhankumar, P., et al.: Fabrication of low-cost MEMS microfluidic devices using metal embossing technique on glass for lab-on-chip applications. J. Micromech. Microeng. **33**(8), 084001 (2023)
104. Huff, M.: Recent advances in reactive ion etching and applications of high-aspect-ratio microfabrication. Micromachines **12**, 991 (2021). https://doi.org/10.3390/mi12080991
105. Koester, D.A., Baumhover, J.P.: MEMS Foundry for High Volume Manufacturing and Product Reliance, pp. 535–540 (2002)
106. Ziaie, B., et al.: Hard and soft micromachining for BioMEMS: review of techniques and examples of applications in microfluidics and drug delivery. Adv. Drug Deliv. Rev. **56**(2), 145–172 (2004)
107. Rahmanian, M., et al.: A micropillar array-based microfluidic chip for label-free separation of circulating tumor cells: the best micropillar geometry? J. Adv. Res. **47**, 105–121 (2023)
108. Sartipzadeh, O., et al.: Microfluidic-assisted synthesis and modeling of stimuli-responsive monodispersed Chitosan microgels for drug delivery applications. Sci. Rep. **12**(1), 8382 (2022)
109. Sartipzadeh, O., et al.: Controllable size and form of droplets in microfluidic-assisted devices: effects of channel geometry and fluid velocity on droplet size. Mater. Sci. Eng. C **109**, 110606 (2020)
110. Sartipzadeh, O., et al.: Microfluidic-assisted synthesis and modelling of monodispersed magnetic nanocomposites for biomedical applications. Nanotechnol. Rev. **9**(1), 1397–1407 (2020)
111. Garshasbi, H.R., Naghib, S.M.: A review on nano/microfluidic devices for cell isolation techniques: recent progress and advances. Curr. Nanosci. **20**(6), 717–739 (2024)

112. Mohammad-Jafari, K., Naghib, S.M.: 3D printing of microfluidic-assisted liposomes production for drug delivery and nanobiomedicine: a review. Curr. Med. Chem. (2024)
113. Naghib, S.M., Mohammad-Jafari, K.: Microfluidics-mediated Liposomal nanoparticles for cancer therapy: recent developments on advanced devices and technologies. Curr. Top. Med. Chem. **24**, 1185–1211 (2024)
114. Shanehband, N., Naghib, S.M.: Microfluidics-Assisted Tumor Cell Separation Approaches for Clinical Applications: An Overview on Emerging Devices. Comb Chem High Throughput Screen (2024)
115. Shanehband, N., Naghib, S.M.: Recent advances in nano/microfluidics-based cell isolation techniques for cancer diagnosis and treatments. Biochimie **220**, 122–143 (2024)
116. Chen, W., Lam, R.H.W., Fu, J.: Photolithographic surface micromachining of polydimethylsiloxane (PDMS). Lab Chip **12**(2), 391–395 (2012)
117. Chen, Y., Zhang, L., Chen, G.: Fabrication, modification, and application of poly(methyl methacrylate) microfluidic chips. Electrophoresis **29**(9), 1801–1814 (2008)
118. Beebe, D.J., et al.: Microfluidic tectonics: a comprehensive construction platform for microfluidic systems. Proc. Natl. Acad. Sci. **97**(25), 13488–13493 (2000)
119. Yu, Q.: Development of functional polymeric materials for microfluidic systems. University of Illinois at Urbana-Champaign (2002)

Chapter 3
BioMEMS and Revolutionizing Medicine: Advances in Microelectromechanical Systems for Effective Drug Delivery

Abstract Microelectromechanical systems (MEMS) have revolutionized medicine, particularly in the field of drug delivery. Traditional methods of drug delivery, such as pills, often fail to effectively target specific areas, but MEMS-based approaches offer innovative solutions. By utilizing micro- and nanoscale drug carriers like micro patches, microparticles, bio-adhesives, and nanoparticles, drug delivery can be significantly improved. Advancements in oral drug delivery have been made possible through the development of these novel delivery methods. MEMS-based systems enable the formulation of drugs in forms that enhance absorption and therapeutic efficacy, optimizing treatment outcomes. Percutaneous transmission, an alternative to oral delivery, also shows promise. Microneedles allow drugs to penetrate the skin barrier without causing pain or nerve stimulation, improving the efficiency of drug delivery. In this chapter, we explore the various drug delivery methods facilitated by BioMEMS and their potential impact. We discuss the utilization of microfabrication techniques and MEMS technologies, enabling precise control over drug release and improved targeting. By examining the application of micro- and nanoscale drug carriers, as well as microneedles, we highlight the transformative potential of BioMEMS in drug delivery. We also address the challenges and opportunities associated with these advanced drug delivery methods. Optimization of drug formulation, carrier design, and delivery mechanisms is crucial for effective and safe administration. Rigorous in vivo testing, biocompatibility assessment, and reliable manufacturing processes are essential to translate these concepts into practical clinical applications. Implantable drug delivery systems have also benefited from BioMEMS technology, offering long-term, controlled drug release tailored to specific patient needs. Ongoing research focuses on improving patient compliance, device size, delivery rates, and biocompatibility. Integration of sensor technology enables real-time monitoring of physiological changes, ensuring precise and personalized medication administration. While challenges remain, including scaling-up processes, long-term performance stability, and material selection, the future of drug delivery lies in advancing microfabrication techniques and BioMEMS. These innovations have the potential to transform personalized medicine, improving patient outcomes and quality of life.

S. M. Naghib and Y. Rezaian, *Frontiers of Nanobiotechnology*, Series in BioEngineering, https://doi.org/10.1007/978-3-031-88419-1_3

Keywords BioMEMS · Biomedicine · Drug delivery · Gene delivery · Nanoparticles

3.1 Introduction

3.1.1 History of Bio-Microelectromechanical Systems (BioMEMS)

BioMEMS (Biological or Biomedical Micro-Electro-Mechanical Systems) are nano or microscaled systems or devices used for analyzing, manipulating, delivering, or processing chemical and biological substances for biomedical or biological purposes. The origins of BioMEMS were documented in the late 1980s and early 1990s, and they now offer a promising new approach to addressing healthcare challenges in the twenty-first century [1, 2]. One notable advantage of BioMEMS is its ability to perform analysis with smaller sample sizes. The manufacturing process enables the designing and creating of small and portable diagnostic kits. Additionally, BioMEMS have demonstrated greater sensitivity and faster analytical times compared to standard laboratory methods. Their size similarity to organs and cells allows for more in-depth research. Importantly, after technological development, BioMEMS can be mass-produced cheaply and with high reproducibility. These inherent advantages have generated significant interest in research within this field [3, 4]. This chapter briefly reviews and analyzes the current characteristics and prospects of microfabricated devices for drug delivery management. It focuses on various administration methods, including implantable systems, and transdermal, oral, and ocular routes. Microfabrication technology holds the potential to offer new and innovative approaches to developing sophisticated and advanced drug delivery systems that can greatly enhance medical treatment.

3.1.2 Why Build BioMEMS?

The development of tiny biomedical systems capable of residing inside the human body or animals represents the pinnacle of future biomedical advancements. These miniaturized systems offer the potential for easier and safer medical procedures, such as endovascular catheterization, injections using large-gauge hypodermic needles, or minimally invasive localized surgeries. Such advancements could significantly simplify the delivery and implantation of these devices [4]. When properly designed, microdevices implanted in the body should have minimal impact on host tissues. Moreover, scaling down the size of a system also scales down the fundamental laws governing its behavior and functionality. This parallels the motivation behind microelectronic engineers shrinking transistors: smaller transistors operate faster and require less power to activate. Similar advantages are anticipated for electromechanical systems, where reducing size enhances their ability to be set in motion or respond rapidly to specific stimuli or environmental changes. In many cases, changes in the surface-to-volume ratio are linked to the scaling of controlling laws and forces. For instance, consider a simple cube shape: as its dimensions (length, thickness, width) decrease, its volume (which is a function of the third power of ordinary dimensions) decreases at a faster rate compared to the total surface area (which is a function of the second power of ordinary dimensions). Similarly, as the size of MEMS systems decreases, heating phenomena are more influenced because cooling (through surface heat exchanges) and heating are impacted by volume exchanges. A notable example of these benefits is the advantageous shrinking of devices like macroscopic di-electrophoresis separators (DEP) [5].

3.2 Advantages of BioMEMS

BioMEMS (Bio-microelectromechanical Systems) has emerged as a crucial technology in the field of drug delivery, offering a wide range of advantages and advancements. These miniature devices enable precise control over drug administration, ensuring accurate dosing and release profiles. By incorporating microscale components and sensors, BioMEMS can manipulate drug release kinetics and respond to specific stimuli like pH, temperature, or biomarkers. This level of control allows for targeted and localized drug delivery to the desired site of action, minimizing systemic toxicity and improving therapeutic efficacy [5]. One of the key benefits of BioMEMS in drug delivery is their ability to facilitate personalized medicine approaches. With integrated sensors and feedback mechanisms, BioMEMS devices can monitor physiological parameters and adjust drug dosages accordingly. This customization enhances treatment outcomes and promotes patient compliance, leading to more effective therapies [6]. Implantable BioMEMS devices offer sustained and controlled drug release, eliminating the need for frequent dosing and enhancing patient convenience. These devices can be designed to release drugs over an extended period, making them suitable for chronic conditions or diseases requiring continuous therapy. The precise control of drug release rates provided by BioMEMS ensures optimal therapeutic effects [7].

Moreover, BioMEMS contribute to the minimization of systemic toxicity. By delivering drugs directly to the target site, BioMEMS bypass other tissues and organs, reducing the potential for off-target effects and adverse reactions. This localized drug delivery approach significantly improves safety and efficiency in drug administration [8]. BioMEMS also offer integration with other technologies, such as microfluidics, nanotechnology, and wireless communication. These integrations enhance drug delivery capabilities, allowing for precise control of fluid flow, drug mixing, real-time monitoring, and remote control of drug delivery processes. The multifunctionality of BioMEMS enables the integration of multiple instruments into a single device, streamlining automation and reducing the need for human interaction [9]. In addition to their role in drug delivery, BioMEMS possess several advantages over traditional techniques. They offer automation, high throughput, replication reliability, portability, and small device dimensions. These features result in lower production costs, reduced sample sizes and reagent usage, and improved mobility. The versatility of BioMEMS makes them suitable for various industries, particularly in medical centers and hospitals [10–14]. The field of BioMEMS is rapidly expanding, with a significant increase in research and market demand. The number of citations related to BioMEMS has been steadily growing, and the global market is expected to exceed $2.45 billion by 2024. This growth is driven by advancements in portable analytics systems, home care devices, pharmacological research, and biological diagnostics. Furthermore, ongoing development in disease-prediction devices holds promise for preventing deadly diseases like HIV or cancer [15–17]. Successful treatment outcomes in drug delivery require the right drug to be transported to the correct site of action at the right time, whether it's a small molecule or a large macromolecule. Each molecule has its unique profile of toxicity (ADMET), excretion, metabolism, distribution, and absorption. The timing of drug release can also impact its efficacy. Therefore, methods that enable precise control over drug distribution have garnered significant interest, with microelectromechanical systems (MEMS) being a focal point in drug delivery research [18–20].

MEMS offers the capacity to fabricate two-dimensional (2-D) and three-dimensional (3-D) material structures with exceptional precision and consistency, presenting significant improvements over traditional drug delivery methods. External drug delivery systems, which do not require implantation, are an essential starting point. These systems include micromachined devices and microneedles designed to enhance buccal, ocular, or transdermal administration. Additionally, microfabricated systems play a role in formulating medicinal preparations for internal use, such as microfluidic mixers and devices for generating injectable or inhalable particles. Formulation devices like gastrointestinal (GI) patches and other devices involved in the formulation process fall under this category [21–23].

For drugs that are challenging to administer orally, devices attached to the intestinal wall provide a solution by facilitating their release. Another group consists of internally implanted devices, including immunoisolation capsules for therapeutic cells, non-Fickian release depots, and micro reservoir systems (e.g., pancreatic islets). The future of MEMS-based drug delivery may rely on implanted microfluidic networks that accurately administer complete treatment regimens. These networks would be supervised by integrated circuits equipped with wireless telemetry and built-in sensors. The progress in microfluidics and labs-on-a-chip technologies is closely intertwined with advancements in the field of MEMS [24, 25]. By merging the fields of microfluidics, MEMS, and drug delivery, researchers are exploring the potential for developing intricate drug delivery systems capable of precisely targeting specific cells, tissues, or organs. These advancements hold great promise for improving treatment outcomes, enhancing patient comfort, and reducing the potential for adverse effects. The continuous progress in these fields paves the way for transformative approaches in personalized medicine and targeted therapies[26].

3.3 Oral Drug Delivery by BioMEMS: Enhancing Efficiency and Precision

Among the various traditional methods of drug delivery, such as invasive injections (intramuscular and intravenous), oral drug delivery is a suitable and highly patient-compliant form of treatment. However, the overall effectiveness of medicines administered orally is limited by certain physiological obstacles in the gastrointestinal (GI) tract [27–29]. One crucial factor for oral treatments is the low gastric pH (approximately 2) in the stomach and the neutral pH (around 7) in the intestines. Gastric acidity is important for digestion and acts as a barrier against infections. However, the low pH poses challenges in delivering drugs as it can disrupt the active ingredients. This becomes problematic for drugs targeting the small intestine because the stomach's primary role in digestion is to break down proteins, making it difficult for enzymes in the small intestine to absorb them as individual amino acids [29]. pH-sensitive compounds that are absorbed in the intestines can only pass through the stomach barrier when they are encapsulated in a non-degradable compound. Particularly for active therapeutics like peptides and proteins, the stomach's acidic environment, combined with various intestinal enzymes, can cause drug decomposition. The tight junctions and adherent mucus of the intestinal epithelium also limit the penetration of high molecular weight drugs into the gastrointestinal tract. Additionally, flow conditions and shear forces resulting from GI tract peristalsis can reduce the retention and bioavailability of medications. Through microfabrication, researchers have developed various delivery devices with unique designs and precision engineering for safe and efficient oral medication administration [30, 31].

Microstructured devices may also improve drug loading efficiency and provide better control over drug release [32–39]. Using multilayered patch systems is one way to improve stability and absorption at the intestinal epithelium. These patches are made up of four layers: an impermeable backing, a drug reservoir, a rate-controlling membrane, and an adhesive. Applying the patch causes the medication to enter the bloodstream via the skin at a regulated rate set by the membrane, which is pre-programmed to preserve the drug's potency. Oral dosage formulations that are intended for intestinal mucosa delivery are extremely attractive due to these features. Three main purposes are served by microfabricated patch systems for oral drug administration: (1) controlled drug release, (2) unidirectional release towards the intestinal epithelium, and (3) bioadhesive properties for dosage form retention [40, 41].

3.4 Micro Patch Technology: Advancements and Fabrication Techniques

Micro patches are specialized drug delivery systems designed for oral administration. They are thin, flat devices that adhere to the mucosal surface of the intestine, aiming to enhance drug absorption and provide controlled release of medications [42]. One of the most important factors in how well oral drugs travel through the digestive system is their size. The protective mucosal barrier generally traps large particles, shortening their residence period. But in Peyer's patches—tiny regions of the brush border that lead to lymphoid tissue specific nanoscale and microscale particles may penetrate the epithelium. Toxic effects and inadequate biodistribution are possible outcomes of this mode of transport [43]. As a substitute for more conventional particle distribution systems like microspheres, micro patches have evolved to meet these difficulties. Intestinal absorption is enhanced by the use of these tiny patches, which may glide effortlessly between the villi of the intestines. At the same time, they are wide enough to prevent cell absorption. Unlike particle systems, micro patches have a thin and smooth structure that maximizes contact with the intestinal surface. Additionally, their flat shape reduces exposure to the fluids flowing in the intestines [44]. The composition of micro patches typically involves multiple layers, each serving a specific purpose. In the case of oral protein administration, researchers have developed three-layer patch devices (Fig. 3.1). These devices sandwich the drug between a sticky polymer layer and an impermeable polymer layer. The sticky polymer layer facilitates the patch's adherence to the intestinal mucosa, while the impermeable layer prevents drug spreading to the external lumen and protects it from enzymatic degradation [45].

Ethylcellulose (EC) is commonly used to create the impermeable dorsal layer, which is less permeable compared to the mucosal adhesive layer. This directs the drug from the mucosal adhesion into the epithelium, ensuring targeted delivery and

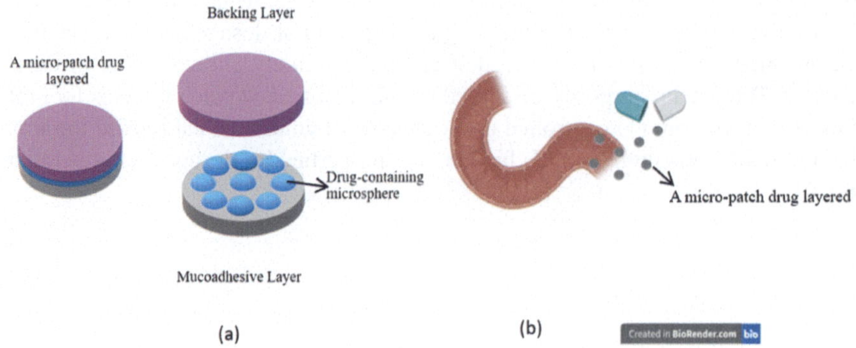

Fig. 3.1 An oral drug delivery system based on a micro patch system, **a** designing a drug-containing patch, **b** Intestinal drug administration using intestinal adhesives. Image created with BioRender.com

preventing exposure to the external environment. Encapsulation of the patches in circular or square forms allows for drug transfer to the intestine. The flat and layered shape of micro patches offers numerous advantages over conventional pills and viscous microspheres. The flat design of the drug delivery system increases the surface area, which improves the adherence to the intestinal wall and prolongs the time of drug release at the desired location. Micropatch fabrication utilizes microfabrication techniques, enabling precise engineering and customization. These devices can be tailored to deliver specific drugs or combinations of drugs, offering personalized treatment options. The incorporation of controlled release mechanisms in micro patches ensures regulated and sustained drug release, optimizing therapeutic efficacy [46, 47]. In conclusion, micro patches provide a promising approach to oral drug delivery. Their unique design, increased surface area, and controlled release capabilities make them an attractive option for enhancing drug absorption, improving bioavailability, and maximizing therapeutic outcomes. By overcoming the limitations of traditional particle delivery methods, micro patches contribute to the advancement of effective and efficient treatment regimens.

Micro Patch Fabrication

Micro patch fabrication is a complex process that involves creating specialized drug delivery systems known as micro patches. These patches are designed to adhere to the mucosal surface of the intestine, providing a controlled release of medications and enhancing drug absorption [48]. The fabrication process of micro patches consists of several key steps. It begins with designing and selecting suitable materials, including determining the shape, size, and composition of the micro patch. Common materials used in micro patch fabrication include polymers such as ethylcellulose (EC) for the impermeable layer and mucoadhesive materials for enhanced adhesion to the intestinal mucosa [49]. Microfabrication techniques are then utilized to create the micro patches with precision. Techniques like soft lithography, thin film deposition, etching, and photolithography are employed to control the dimensions and features of the patches. These techniques enable the creation of microscale structures within the patches [50].

The next step involves drug loading, where specific drugs or combinations of drugs are incorporated into the micro patches. This is achieved by filling the reservoirs within the patches with the desired medications or biomolecules. Drug loading techniques include microinjection or using a drug solution in a polymeric material that solidifies within the reservoir [51]. Micro patches are designed to provide a controlled release of drugs over a specific period. Various mechanisms are incorporated into the patch design to regulate the release of the drug. For example, hydrogels that respond to temperature, solvent, or pH changes can control the release rate. Additionally, different polymers with varying dissolution rates can be utilized to achieve controlled release of different compounds [52].

To prevent medication leakage from the side edges of the patch, specific patches can be coated with materials of limited permeability. For instance, a 400-micron-thick disk of sodium carboxymethylcellulose, pectin, and carboxyl can serve as a mucoadhesive matrix to distribute the medication, with an ethylcellulose (EC) layer covering all but one side of the disk. This protective covering has been shown to reduce drug loss and increase bioavailability in porcine models [53]. Surface modifications can be applied to the micro patches to include specific features or targeting mechanisms. These modifications help direct the patches to specific action sites within the intestine. Incorporating cell-targeting ligands or functionalizing the surface of the patches enhances their interaction with specific cells or tissues [54]. After fabrication, the micro patches undergo testing and evaluation to assess their performance. This includes studying drug release profiles, stability, adhesion properties, and bioavailability. In vitro and in vivo experiments may be conducted to validate the effectiveness of the micro patches [55]. Micro-patch fabrication requires expertise in microfabrication techniques, material science, and drug delivery principles. By combining precise engineering, controlled release mechanisms, and targeted drug delivery, micro patches offer a promising approach to enhancing oral drug absorption and improving therapeutic outcomes. They provide advantages such as precise drug loading, predetermined release profiles, and improved drug stability, making them a valuable tool in the field of drug delivery [56].

3.5 Bio-adhesion in Drug Delivery: Leveraging BioMEMS for Enhanced Adhesion

Bio-adhesion occurs through close interaction between drugs and mucosa, and it can take different forms and involve various types of mucous membranes. Mucous membranes line the walls of the digestive tract, inner eyelids, nasal and oral cavities, and genitals, serving as gateways for substances while protecting against harmful ones (Fig. 3.2) [57]. For example, bio-adhesives address challenges related to short gut residency durations and insufficient mucosal contact in drug delivery and biomedical systems [58]. These innovative systems are designed to prolong the residence time of medications at specific delivery locations, thereby improving medication absorption and safeguarding the drugs during absorption [59]. Bio-adhesion may be accomplished by either physical or chemical means. Physical approaches use adhesion triggers, such as tethered or linear polymer chains, to augment the adherence of drugs during oral delivery. Chemical approaches, in contrast, use the composition of artificial polymers to enhance mucoadhesion in controlled-release applications. Immobile lectins have been used as targeted agents to increase adhesion and interaction with intestinal epithelial models. Microdevices, such as micro-patches and microneedles, have been modified to exhibit bio-adhesive properties. These modifications include chemical alterations of the device's surface to immobilize bioactive targeted molecules. In vitro studies using lectin-conjugated microdevices have demonstrated their cytoadhesive characteristics when interacting with Caco-2 cells, a widely used cell model for drug absorption studies. Comparisons between microdevices and microspheres have revealed that microdevices exhibit greater adhesion

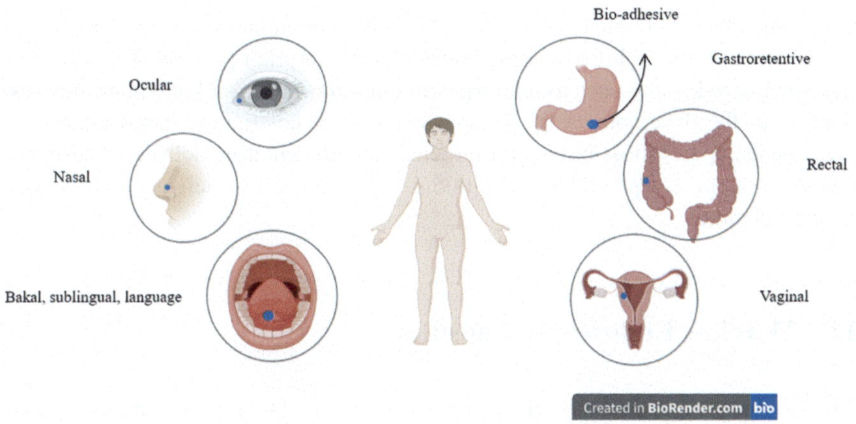

Fig. 3.2 Bio-adhesive drug delivery routes in different organs. Image created with BioRender.com

stability, likely attributed to their larger lectin-modified surface area and increased contact with the cell monolayer [60].

Microneedles, initially developed to improve drug permeability through the skin, have proven effective in enhancing mucosal adhesion. The microneedles are specifically engineered to have a length that allows them to pierce the outermost layer of the skin, known as the stratum corneum, to administer drugs. However, they are also meant to be short enough to prevent any irritation to the nerves. Microneedles may enhance medicine permeability by enhancing the adhesion of particles to the mucosa [61]. Advancements in microfabrication techniques have enabled the creation of oral microdevices with precisely formed microneedles or micro-posts. These microdevices exhibit the potential to penetrate the mucus layer, anchoring themselves and prolonging their retention within the gastrointestinal tract. For example, researchers have developed microdevices with self-folding arms made of specific polymers and crosslinked chitosan microparticles. These arms penetrate the mucus layer, enhancing resistance to erosion and allowing the device to "grip" the intestinal villi. Additionally, the presence of micro-posts on oral microdevices can facilitate the shedding of the mucosa, thereby increasing chemical absorption into the underlying blood arteries. The combination of micro-posts and targeted drug delivery approach offers the potential for mechanically controlled release [62]. Furthermore, bio-adhesion can be enhanced through gecko-inspired or nanostructure-based adhesives. By constructing nanomaterials with increased surface area-to-volume ratios, adhesive components attached to microvilli-coated intestinal epithelium can generate substantial bio-adhesive forces. Nanowire-coated devices, for example, have demonstrated significantly higher adhesion to Caco-2 cells compared to non-coated devices. The geometric characteristics of these nanomaterials contribute to their enhanced bio-adhesive properties [63]. In summary, bio-adhesives provide a promising solution for improving drug delivery and biomedical systems. By prolonging residence time and enhancing adhesion to mucosal surfaces, bio-adhesives enhance medication absorption and ensure the stability of drugs during the absorption process. The advancements in microdevices, such as micro-patches, microneedles, and oral microdevices, along with the development of gecko-inspired or nanostructure-based adhesives, have opened up new possibilities for targeted and efficient drug delivery. These bio-adhesive systems can potentially revolutionize the field of drug delivery and improve patient outcomes.

3.6 Wireless Endoscopic Capsules

Wireless endoscopic capsules offer a noninvasive method for remotely treating disorders of the gastrointestinal (GI) tract, reducing patient suffering [64]. However, constructing a remotely operated drug delivery system (DDS) for capsule endoscopy is challenging due to the restricted and sensitive environment of live tissue. The size of the DDS is limited as it needs to be swallowed (Fig. 3.3). The InteliSite, Intell-iCap, and Enterion capsules are capable of releasing medicines, with a capacity of up to 1 mL [65]. Among these, InteliSite is the most popular remote-control device for localized medication absorption. It provides various dose forms, including small tablets, particulate matter, suspensions, and solutions. Although DDS for capsule endoscopy has advanced, the drug release from these capsules is not fully regulated and they do not have an anchoring mechanism. To get over these restrictions, newer versions of capsule endoscopes combine several techniques to deliver medications at certain GI tract locations [66].

Some studies have examined the development of untethered release mechanisms using MEMS technology, however much of the research has mostly focused on non-mechanical methods. MEMS-based capsule endoscopy devices commonly employ tiny batteries to activate and power the drug-releasing mechanism within the capsules. Some trials have also used anchoring methods to securely adhere the endoscopic capsules to the GI tract walls before releasing the medication [67, 68]. However, these processes require a significant amount of power, which is not readily supplied by conventional batteries. As a result, researchers have explored remote charging and actuating of MEMS-based devices. Swallowable wireless endoscopic capsules, such as those used to examine the GI system and administer targeted therapeutics, have been developed. Commercial capsule endoscope prototypes like Enterion capsule, Pulsincap, InteliSite, and IntelliCap have been created to achieve triggered medication release using various methods [69].

These devices usually consist of an enclosed assembly comprising a medication reservoir, a piston, and a polymer capsule with an energy producer-transmission system. Using either the energy from the gadget or an external stimulus, the piston moves and releases the medication. For instance, an InteliSite capsule measuring 35 mm in length and 10 mm in diameter is placed next to a compressed spring that has two alloy wire clamps, and it contains a 0.8 mL medication tank. The wire clamps give way after two minutes of antenna activation, letting the spring exit the capsule

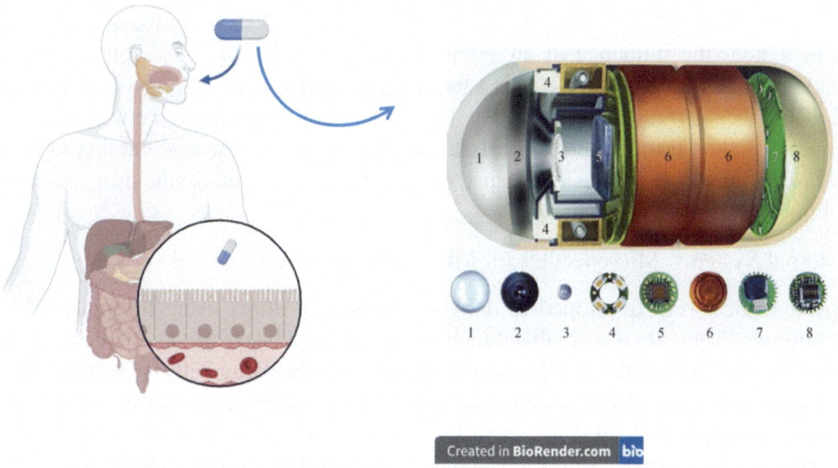

Fig. 3.3 **a** endoscopy procedure. Image created with BioRender.com, the image **b** showcases the endoscopic capsule, consisting of several components. The capsule is depicted with its key elements illustrated in the schematic diagram. These components include an optical dome (1), lens holder (2), a short focal length lens (3), four LEDs (4), a CMOS image sensor (5), two silver oxide batteries (6), an ASIC radio-frequency transmitter (7), and an external receiving antenna (8). Reprinted with permission from Ref. [77]

and release the medication [69]. To move the piston and push the medication out of the reservoir, some capsule endoscopes use a pressed stretchable component or a thermosensitive spring composed of shape memory alloy. Gas pressure can generate piston motion and operate the spring-like mechanics. Some examples include gas generation through electrolysis or the use of solid propellant micro thrusters ignited by an external radio frequency signal [70]. Most current capsule endoscopes rely on passive movement driven by the GI tract's natural peristalsis, which limits control over their orientation, location, and speed of movement. Active and controlled movement methods have been developed, such as a legged mechanism with brushless DC motors driven by a battery. Some legs of the capsule resist peristaltic action in parts of the digestive system like the gut [71]. Anchoring systems and needle positioning mechanisms have been incorporated into capsules to administer medicine specifically to the GI tract. Controlled medication release in the GI tract can be achieved using capsule endoscopes as described above. However, due to the limited space within the capsules, including mechanical/electronic systems and onboard batteries, there is minimal room for medication loading [72, 73]. To enhance medication loading and localization capabilities, a capsule device relying solely on magnetic effects was proposed. It consists of two metallic pieces that magnetically link and keep the capsule closed as it travels through the GI system. An external magnetic field is used to demagnetize the metallic components and release the medicines. This device has a larger drug reservoir than similar-sized devices, but medication is only released once in a specific area [74, 75]. To administer multiple doses in a controlled manner, a prototype of a magnetically activated soft capsule was developed. The capsule has two permanent magnets on either end, with the drug-filled chamber located between them. Under the influence of an external magnetic field, the moveable magnetic head squeezes the chamber, causing the drug to be delivered through its fractures. The chamber can revert to its initial shape when the external magnetic field is no longer present. Further improvements are needed to enhance the repeatability of drug release, miniaturize the device, reduce drug leakage, and increase the drug chamber volume [76].

External Systems: Microneedles for Minimally-Invasive Drug Delivery

Microneedles were developed to deliver transdermal drugs using foreign MEMS technology. They are a specialized technology designed to facilitate the delivery of drugs into the body. Microneedles can be used for various drug delivery applications, including transdermal delivery, intradermal delivery, and targeted delivery to specific tissues or regions [78, 79].

Microneedles offer several advantages over traditional drug delivery methods, such as hypodermic needles [80–83]:

1. Minimally Invasive: Traditional transdermal medication delivery is significantly hampered due to the stratum corneum, which is a layer of dead skin tissue that is 10–20 mm thick [9]. In contrast, the skin's nerve endings are positioned a few hundred microns below the skin's surface. While needles are the easiest way to break down skin barriers, traditional needles are uncomfortable and may lead to infection or scarring of underlying tissues [6]. Microneedles, on the other hand, are minimally invasive compared to hypodermic needles, as they create tiny micrometer-scale channels in the skin instead of penetrating deep into the tissue. This reduces pain, discomfort, and the risk of injury or bleeding.
2. Drug Delivery via the Skin: Microneedles are often used for drug delivery via the skin, which involves introducing medications into the systemic circulation. By slipping past the stratum corneum the skin's outermost layer and a natural barrier to medication absorption, they improve drug penetration. This method may be used to avoid the pain of intravenous injection, the early hepatic effects linked to oral delivery, and the breakdown of digesting molecules. Proteins and other medications with low oral bioavailability have seen increased use of this technique in recent years.
3. Targeted Delivery: Microneedles can be designed to target specific tissues or regions within the body. For example, intradermal microneedles are shorter and are used for precise delivery into the dermal layer of the skin. This is particularly useful for vaccines, allergen immunotherapy, and local anesthesia. In the future, longer and stronger microneedles may be developed to effectively penetrate the skin's layer while still being short enough to prevent activating nerves. Microneedles of roughly 100 μm in length can enter the barrier into the underlying interstitial fluid and cells without producing discomfort.
4. Controlled Drug Release: Microneedles enable controlled release of drugs. They can be coated or loaded with drug formulations, allowing for precise dosing and

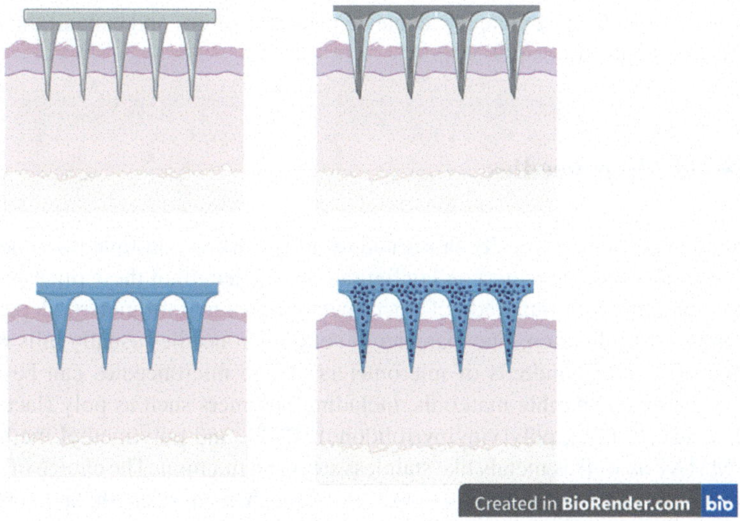

Fig. 3.4 Delivery of dermal medications with microneedles; (1) Solid microneedles are removed after use and replaced with a transdermal patch; (2) The surface of the microneedles is covered with a drug formulation; (3) microneedles are made using a soluble drug formula; (4) The drug solution is injected with a hollow microneedle. Image created with BioRender.com

release kinetics. Some microneedles are designed to dissolve or degrade over time, releasing the drug gradually.

5. Patch-Based Systems: Microneedles are often integrated into patch-based systems. These patches contain an array of microneedles that deliver drugs when applied to the skin. This approach offers convenience, ease of use, and controlled drug release.

Four different methods for transdermal medication administration have been identified (Fig. 3.4) [84–88]:

1. Hollow Microneedles: This method involves using hollow microneedles to inject vaccines or medicine directly into the skin.
2. Dissolving Microneedles: With this approach, dissolving microneedles are applied to the skin, releasing medicine as they dissolve and penetrate the skin.
3. Coated Microneedle Array: In this method, the microneedle array is coated with a medication formulation before puncturing the skin, allowing for controlled drug delivery.
4. Poke and Patch Technique: This technique involves removing the microneedle array after puncturing the skin and replacing it with a transdermal patch that covers the perforated area, facilitating continuous drug release.

Therefore, microneedles as a medical system for drug delivery is an active area of research and development. Ongoing advancements in microneedle technology aim to improve drug delivery efficiency, enhance patient comfort, and expand the range of therapeutic applications [89, 90].

3.7 Solid Microneedles

Solid microneedles are typically fabricated as arrays, where multiple microneedles are arranged in a specific pattern or configuration. The length of these microneedles can vary, ranging from hundreds of micrometers to a few millimeters, depending on the intended application. The diameter of solid microneedles usually falls within the range of tens to hundreds of micrometers. These microneedles can be made from various biocompatible materials, including polymers such as poly (lactic-co-glycolic acid) (PLGA), poly(vinylpyrrolidone) (PVP), and poly(methyl methacrylate) (PMMA), as well as metals like stainless steel and titanium. The choice of material depends on factors such as biocompatibility, mechanical strength, and compatibility with the manufacturing process [91]. Solid microneedles facilitate drug delivery through various mechanisms. They utilize passive diffusion to transport medication into the epidermal layers, enabling both systemic and local effects. It's been shown in experiments using confocal laser scanning microscopy that transdermal distribution of medicine is significantly reduced by skin's inherent elastic properties and the rapid mending of micro-perforations within 15 min. Occlusive covering (like a patch) can help keep pores open for extended periods. Researchers have explored different fabrication methods and materials for solid microneedles. For instance, the tetramethylammonium hydroxide etching method has been employed to manufacture solid silicon microneedles with specific dimensions, including a base width of 110.5 μm and a height of 158 μm. Additionally, gold-coated solid silicon microneedles with specific dimensions and angles have been developed, resulting in improved mechanical strength and bioavailability [92–94].

Studies have demonstrated that biodegradable polymer solid microneedles, such as those made from polylactic acid, possess sufficient mechanical strength to penetrate the stratum corneum, which is the outermost layer of the skin, thereby enhancing medication absorption. Researchers have successfully increased drug penetration by increasing the depth and density of the needles. Stainless steel microneedles have also been investigated, showing promising outcomes for the administration of specific drugs such as captopril and metoprolol tartrate [95, 96].

3.8 Coated Microneedles

Coated microneedles are a variation of microneedles that have a coating applied to their surfaces. This coating serves different purposes, including enhancing drug delivery, improving stability, and modulating the release of drugs or other substances

[97, 98]. Coating tiny needles with a drug and inserting them allows for an effective treatment method. The "coat and poke" approach is used when the required medication quantity is very small or the substance is highly potent. The drug solution or dispersion layer surrounds the microneedles, and later, the coating dissolves, enabling rapid delivery. The amount of medication loaded varies depending on the thickness of the coating layer and the needle size [99, 100]. Coatings on microneedles can be designed to enhance the penetration of drugs through the skin. They modify the surface properties of the microneedles, reducing their interaction with the skin's stratum corneum and facilitating the transport of drugs into the deeper layers. This improves the effectiveness and efficiency of transdermal drug delivery [101]. In addition to drug delivery, coatings can improve the stability of microneedles. Certain coatings protect the microneedles from degradation or chemical reactions with the drug formulation. This is particularly crucial for sensitive drugs or situations requiring long-term stability [102].

Various coating formulations employ microparticles, proteins, viruses, and DNA. This method is particularly suitable for vaccine administration as the required vaccine dosage is often modest, and a shallow penetration depth is sufficient to engage antigen-presenting dermal dendritic cells and epidermal Langerhans cells. Research has explored the use of microneedle arrays to enhance immunization efficacy, with studies showing that increasing needle length beyond 200 microns does not further improve vaccination effectiveness. Coated microneedles have been used for vaccines against viruses such as Chikungunya, West Nile, and influenza, and for hepatitis C DNA immunization [102, 103]. Coated microneedles have also been investigated for the delivery of other medications, including genetic materials, the hormone desmopressin, and parathyroid hormone. One significant advantage of coated microneedles is the ability to stabilize drugs at room temperature for extended periods. Solid coatings on microneedles maintain drug stability, enabling the shipment of vaccines and medications globally, including to low-resource areas lacking cold storage facilities [104]. Various additives, such as trehalose, have been used to stabilize vaccines and prolong their shelf life at room temperature. Trehalose-stabilized vaccines can be incorporated into coated microneedles, and the addition of a porous covering like calcium phosphate further enhances their stability [105]. Researchers have also explored specific drug coatings on microneedles. For example, lidocaine was loaded onto poly L-lactide (PLLA) microneedle arrays, demonstrating rapid release in phosphate buffer saline. Investigations are underway to develop coated microneedles capable of delivering multiple drugs simultaneously using different formulations. This approach allows for the delivery of drugs with varied characteristics, including both water-soluble and water-insoluble substances. Studies have shown high drug delivery effectiveness, with continuous release verified in in vitro tests [106]. Furthermore, coated microneedles enable the controlled release of drugs. The coating can act as a reservoir or a barrier, controlling the release rate and duration of drug delivery. This can be achieved through the selection of appropriate coating materials or the incorporation of drug-loaded particles within the coating [106].

In conclusion, coated microneedles provide a versatile and efficient method for delivering drugs and vaccines. The combination of microneedle technology with drug coatings offers advantages such as enhanced drug stability, controlled release, and the ability to deliver multiple substances simultaneously. Ongoing research aims to optimize coating formulations, stability, and delivery efficiency, further expanding the applications of coated microneedles in the field of drug delivery.

3.9 Dissolving Microneedles

Dissolving microneedles, also known as biodegradable or soluble microneedles, offer a convenient and painless method for drug delivery. These microneedles are designed to penetrate the skin and dissolve, releasing the encapsulated drug or vaccine into the targeted tissue [107]. Dissolving polymer matrices filled with drug formulations can be used to make entire needles, optimizing drug loading and reducing the harmful consequences of fragmented microneedle pieces left in the skin. This maximizes the medication loading volume and avoids hazardous biohazardous waste. However, the fabrication process is more complex and may include processes that could harm the medication [108]. Dissolving microneedles are typically made from biocompatible and biodegradable materials, such as polymers, sugars, or proteins. These materials are carefully selected to ensure that the microneedles maintain their structural integrity during penetration but dissolve quickly upon insertion into the skin [109]. The medication is released when the microneedle is inserted into the skin and dissolves. Unlike other situations where the microneedle is withdrawn after insertion, this procedure requires only one step. Inside the skin, the polymer degrades, regulating the release of the medication. The dissolution of the polymer and its bio-acceptability within the skin make dissolving microneedles an excellent option for long-term treatment with improved patient compliance [108]. Early advancements in dissolving microneedles involved materials such as maltose, polyglycolic acid (PGA), polylactic acid (PLA), and their copolymers, which are gradually biodegradable solid needles used in the "poke-and-patch" technique. However, integrating unstable medicinal ingredients during needle casting without compromising drug activity has been challenging due to the high melting temperatures of these biodegradable polymers [43]. To address this issue, manufacturing techniques have been developed that utilize lower temperatures. Microparticles containing the drug can be placed into needle molds and melted using ultrasonic or lower (but still high) temperatures. This method allows for the creation of a layered construction with different materials along the needle's length. Recently, carboxymethylcellulose needles have been manufactured using a multistep process that involves casting and drying the needles first, followed by casting and drying the base on top. Importantly, all of these processes are completed at body temperature, preserving the action of the encapsulated medication even after extended dry storage [97]. Furthermore, researchers have made various advancements in dissolving microneedles to improve drug delivery efficiency and control release kinetics. For example, tip-dissolving microneedles have been developed to deliver drugs quickly and effectively without causing skin discomfort. Microneedles that dissolve slowly and have full insertion difficulty have been enhanced by placing them on solid microneedles, providing mechanical strength and achieving 90% delivery efficiency in 30s. Other modifications, such as adding bubbles or creating microneedles with separable arrowhead tips, have demonstrated the potential for fast medication administration with controlled release kinetics [110]. While dissolving microneedles offers advantages, including improved patient compliance and targeted drug delivery, there are also some drawbacks. One notable drawback is the potential precipitation of chemicals or unwanted polymers in the body. Ongoing research aims to address these challenges and further optimize the design, fabrication, and performance of dissolving microneedles for safe and effective drug delivery.

Dissolving microneedles offers several advantages over traditional injections or other drug delivery methods. These include [44, 82, 83, 90]:

1. Painless administration: Dissolving microneedles are designed to be painless, as they do not reach the nerve endings responsible for pain sensation.
2. Improved patient compliance: The painless and minimally invasive nature of dissolving microneedles can improve patient acceptance and compliance with treatment or vaccination regimens.
3. Precise and targeted delivery: The microneedles can be designed to deliver drugs or vaccines to specific tissues or layers of the skin, allowing for precise targeting and minimizing systemic exposure.
4. Enhanced stability: The encapsulation of drugs or vaccines within the microneedles can protect them from degradation or denaturation, improving their stability during storage and transportation.
5. Simplified disposal: Since the microneedles dissolve completely, there is no need for needle disposal, reducing the risk of needlestick injuries and the need for sharps containers.

Dissolving microneedles have shown promise in various applications, including vaccine administration, transdermal drug delivery, and gene therapy. They have been used for immunizations against influenza, measles, and polio. Additionally, dissolving microneedles has been explored to deliver therapeutics, including insulin for diabetes treatment and anticancer drugs. Overall, dissolving microneedles represents an innovative and patient-friendly approach to drug delivery. Ongoing research aims to optimize their design, fabrication techniques, and drug-loading capabilities to expand their applications and improve their effectiveness.

3.10 In Vivo Reliability of BioMEMS: Ensuring Efficacy and Safety

In vivo reliability refers to the ability of BioMEMS (Biological Microelectromechanical Systems) and MEMS (Microelectromechanical Systems) devices to perform reliably and safely when implanted or used inside the human body. These devices are designed to interact with the biological system, providing functionalities such as monitoring, sensing, drug delivery, or tissue engineering [111]. One crucial aspect of in vivo reliability is the prevention or reduction of any short- or long-term harm to the human host. This includes mitigating risks associated with chemical contamination, such as the release of harmful substances from the microdevice (outgassing) or corrosion of materials within the device. These issues must be carefully addressed during the design and fabrication process to ensure the safety of the patient [112, 113]. Mechanical debris is another concern in the context of in vivo reliability. The microdevices should be designed and manufactured in a way that minimizes the generation of debris, as loose particles can cause tissue irritation, inflammation, or even blockages in blood vessels or other biological pathways [114]. The host biological system's natural reaction to the presence of the microdevice is also an important consideration. When the device is implanted, it interacts with the surrounding tissues and fluids. It is crucial for the device to be biocompatible, meaning it should not elicit adverse reactions, immune responses, or tissue damage. Extensive testing and evaluation are conducted to ensure that the microdevice functions harmoniously within the biological environment. Unlike external devices that can be easily removed or replaced, implanted micro-devices often need to remain in the body for an extended period. Therefore, they must exhibit reliability and longevity. These devices are expected to operate according to their specified functionality for the intended duration without experiencing significant performance degradation or failure. This requirement includes maintaining accuracy, precision, and stability over time, as well as ensuring an acceptable level of drift in their performance characteristics [115, 116]. Overall, achieving in vivo reliability requires careful consideration of materials, fabrication processes, biocompatibility, functionality, and long-term performance of BioMEMS and MEMS devices. Extensive testing, evaluation, and validation are conducted to ensure their safety and efficacy when used inside the human body or other critical applications.

3.11 Implantable Drug Delivery Systems

Implantable BioMEMS devices offer a wide range of applications in the field of drug delivery, including drug-releasing implants. These implants have the potential to improve medication administration by providing sustained release over an extended period. One of the key advantages is enhanced drug compliance, as patients do not need to consciously administer the medication regularly. This approach also reduces the cost associated with frequent injections or the need for specialized medical expertise. Implants are especially beneficial for drugs that have low bioavailability when administered orally or through barriers like the vaginal wall, skin, or nasal mucosa. Patients requiring long-term therapies, such as daily or frequent drug injections, can benefit from these implants. However, there are some disadvantages to consider. Implantation of the devices typically requires a simple outpatient surgery, and additional procedures may be necessary to remove, replace, or refill the reservoir with medication. Although these procedures may be considered uncomfortable or disruptive compared to routine maintenance procedures, they could be advantageous depending on the specific disease being addressed. The presence of the implant mustn't trigger an aggressive immune response in the body. To mitigate the risk of dosage leakage, which can be a major failure mode for implanted medical devices, targeted medication release and exposure can be achieved after implantation. This approach minimizes the likelihood of dosage dumping, which is particularly critical for medications like insulin that can be lethal if administered inappropriately. By distributing the medication into numerous individually sealed micro reservoirs, the risk of dosage dumping can be significantly reduced. This method eliminates the need for a separate metering device and allows for the storage of multiple drugs in separate reservoirs within a single implant. To administer the dosage, the device simply needs to open the appropriate reservoir and release the medication. MEMS technology enables the fabrication of nanoliter reservoirs, and electronic controllers can easily manage the reservoir array, opening each one as needed [35, 117, 118]. In summary, implantable BioMEMS devices, including drug-releasing implants, have the potential to revolutionize drug delivery by providing sustained release and targeted administration. These devices offer advantages such as improved drug compliance, reduced costs, and the ability to address drugs with low bioavailability. While there are considerations regarding implantation procedures and the avoidance of adverse immune responses, the use of micro reservoirs and MEMS technology allows for precise dosage control and improved safety.

Types of implant drug delivery systems [119–122]:

(1) Passive implants are generally single and homogeneous devices consisting of a simple medicine package embedded in a biocompatible matrix or substance. These passive drug delivery systems do not have actuators and are regulated by a feedback system. They typically use reservoir-type carriers to release drugs, primarily through drug dispersion or degradation of the delivery device. In a passive delivery system, the medication is usually contained in a dissolving area or a semipermeable, permeable membrane. This results in a predetermined drug release profile, where the timing of drug release is primarily determined by the physical characteristics of the drug (such as permeability and diffusivity) and the properties of the matrix or membrane (including distribution, pore sizes, dissolution rate, etc.). The delivery kinetics can be partly adjusted by selecting the drug, surface properties, matrix material, overall implant morphology, and concentration. Passive drug delivery devices are economical because they may be made to produce intricate pulsatile release patterns without a power supply. Their manufacturing complexity is lowered by the lack of an actuation mechanism. Although their efficiency may be lower compared to active systems, there are still significant advantages to utilizing passive drug delivery devices.

(2) Time-controlled implants offer significant benefits, particularly when loaded with low-diffusion drugs. Their small diffusion volume at a microscopic level allows for the use of high-potency drugs. The release of the drug can be designed and controlled based on the erosion rate of the implant's reservoirs. This is achieved by using coated materials with varying erosion rates, which can be

tailored to specific clinical conditions. For instance, changes in body temperature or the pH of the interstitial fluid can facilitate the accelerated release of the next dose. However, it's crucial to consider that these effects can also have potential drawbacks. They may lead to the delayed release of a necessary dosage or an undesired early release, impacting the effectiveness of the treatment. Nonetheless, time-controlled implants present an innovative approach to drug delivery, offering precise and controlled release of medications over specific time intervals.

(3) User-controlled implants are becoming increasingly sophisticated in both their design and functionality. Although they may not provide the same level of convenience as fully automated implants, they offer significant advantages by reducing the effort required to obtain medication, particularly when it comes to injectable drugs. Many patients prefer the option of pressing a button once a day over undergoing daily injections. Furthermore, user-controlled implants help minimize potential harm to delicate areas, such as the eyes, as they only require a single insertion site instead of multiple injections. These advanced implants incorporate user-controlled features, enabling patients to actively participate in their treatment and medication administration. They can be connected to external devices, such as handheld or wearable devices, which allow patients to initiate the release of drugs or adjust treatment parameters. Additionally, some implants utilize wireless communication technology, allowing patients to wirelessly activate and control the implant's functionality using a smartphone or other compatible devices. While user-controlled implants do require some level of user intervention, the benefits of personalized treatment, reduced injections, and minimized damage to sensitive areas make them a valuable option for many patients. Continuous advancements in this field hold great promise for further improving the ease and convenience of using these implants in the future.

(4) Implants with condition-sensitive capabilities, particularly those designed for active drug delivery, represent ambitious and advanced designs. Unlike passive drug delivery devices, active drug delivery systems incorporate separate power sources to control the opening and closing of drug reservoirs upon request. These systems utilize micro-pumps to distribute medications and can be activated through mechanical or non-mechanical means. The use of active drug delivery systems allows for precise regulation of medication release, enabling customization for the treatment of various diseases. By integrating biosensors and electronic control chips, active medication delivery systems can respond to the body's environment. These implants release carefully measured doses of medication in response to changes in body chemistry detected by the biosensors, such as an increase in blood glucose levels. For instance, implants that monitor blood sugar levels and deliver insulin as needed, mimicking the function of pancreatic islet cells, can greatly benefit patients with diabetes. The development of an "artificial pancreas" has been a long-standing goal. MEMS-based (Micro-Electro-Mechanical Systems) devices play a crucial role in medication delivery. They may be used to administer a single drug or a mixture of medications, depending on the number of reservoirs included. Further, these devices can

deliver medicine in either a continuous or pulsatile fashion. MEMS-based drug delivery systems have demonstrated effectiveness in various therapeutic applications, including central nervous system drug delivery, vaccine administration, diabetes management, treatment of ophthalmic diseases, hemorrhagic shocks, and osteoporosis. It is important to note that these implants are highly complex, and many measurement systems are still in the experimental stage. While they remain a theoretical design goal, significant advancements have been made in their development. Continued research and innovation are required to refine these systems, optimize their size, and ensure their practical implementation in clinical settings. Nonetheless, the potential of implants with condition-sensitive capabilities in revolutionizing drug delivery and personalized medicine holds great promise for the future.

(5) Immuno-isolating biocapsules play a crucial role in protecting implanted cells from the host's immune system, thereby offering a solution for cell-based therapies. These biocapsules exhibit characteristics that combine both passive and active drug delivery systems. Similar to active drug delivery systems, the amount of medication delivered, such as insulin, is determined by the patient's biomarker levels. However, unlike active drug delivery systems, immuno-isolating biocapsules lack active components such as microvalves. In the field of ophthalmology, delivering drugs to the eye's tissues poses significant challenges due to the complex anatomical and physiological makeup of the eye. Various barriers, including the conjunctiva, sclera, cornea, blood ocular barrier (blood-aqueous and blood-retinal barriers), and tears, hinder the effective delivery of drugs to the retina and ciliary body. Current ocular drug delivery methods, such as intraocular injections, oral medications, and topical eye drops, have limited effectiveness, as recent studies have shown. For the treatment of anterior eye diseases, eye ointments, and eye drops are commonly used. Eye drops, which constitute the majority of ophthalmic drugs, have low intraocular bioavailability (around 5%) due to rapid dilution by tears and the need for frequent administration. Efforts have been made to enhance intraocular bioavailability (approximately 10%) by incorporating gels, ointments, and liposomes into eye drops. However, drug administration through oral routes requires high systemic dosages to achieve therapeutic levels in the eye, often leading to serious adverse effects. Furthermore, the poor corneal permeability, clearance through the nasolacrimal ducts, and constant turnover of tears contribute to less than 5% of the dosage being absorbed. To achieve therapeutically effective doses, the prescription frequency or drug concentration must be increased. While the anterior segment of the eye is relatively accessible, targeting it can still be challenging. On the other hand, targeting the posterior segment, which is associated with most vision-threatening ocular disorders, is even more difficult due to anatomical inaccessibility and the blood-retinal barrier. In this context, implanted medication delivery devices offer a promising solution for treating diseases in the posterior segment of the eye. Intraocular implants can penetrate or bypass barriers, thereby increasing the effectiveness of drug delivery. Possible implantation sites include the intraocular space, peribulbar region, pars plana, and vitreous space. Ocular

illnesses including proliferative vitreoretinopathy (PVR) and cytomegalovirus (CMV) retinitis that need procedures and frequent medication injections are ideal candidates for implantable intraocular drug delivery systems. Moreover, these systems are suitable for treating chronic diseases with limited therapeutic options, including retinitis pigmentosa, macular edema, and age-related macular degeneration (AMD). Implanted intraocular drug delivery devices enable the administration of a wide range of medications, including biomolecules and potent antiangiogenic agents, due to localized distribution. Preclinical testing of these implanted devices has been conducted on animals, demonstrating their potential for future clinical applications.

In summary, immuno-isolating bio-capsules provide a protective environment for implanted cells, shielding them from the immune system. In the field of ophthalmology, where effective drug delivery to ocular tissues is challenging, implanted medication delivery devices offer a promising approach for treating anterior and posterior segment eye diseases. These devices have the potential to enhance drug bioavailability, reduce the need for frequent administrations, and provide localized drug distribution, thereby improving therapeutic outcomes and patient convenience in the treatment of various ocular conditions.

3.12 Ocular Drug Delivery via MEMS Technology: Advancements and Challenges

The use of a microelectromechanical systems (MEMS) drug delivery device for treating chronic and refractory eye disorders is currently under investigation. This device is designed to overcome limitations in traditional medication delivery techniques. By employing MEMS technology, the device can meet the size requirements of the ocular orbit (2 mm thickness), and certified biocompatible materials can be used for its assembly. This device offers targeted medication delivery to affected tissues, bypassing physiological barriers, as well as a mechanism to prevent overdosage or unintentional dosage. The compatibility of the drug delivery method and reservoir allows for a broader range of combination treatments and new pharmacological solutions. To enhance patient compliance and reduce associated risks, refillable devices that require only one surgical operation are necessary. These qualities, along with increased flexibility, raise the potential for better efficacy compared to current methods. The MEMS device allows for replenishment of the medication solution, enabling long-term drug therapy without the need for surgery. The first generation of MEMS devices is manually controlled, which limits the duration of drug release and leads to variations in reservoir barrier force. To address these issues, the next generation of devices includes a drug reservoir with a refill gate, an electrolysis chamber with electrolysis actuation for precise dose administration, a battery, and electronics. The MEMS device is constructed using perylene, a biocompatible and flexible material. Electrolysis is driven by batteries and wireless inductive power transmission.

During electrolysis, water is electrochemically converted to oxygen and hydrogen gas, creating pressure in the reservoir that pushes the medication through the cannula. The reservoir is placed in the subconjunctival region, and the flexible cannula is inserted through an incision in either the frontal or back section. Another refillable device based on MEMS technology is an active medication delivery pump. A basic method involves using a manually driven mechanical system to expel the pharmaceutical solution from a storage container via a small tube. The reservoir is inserted into the subconjunctival region with the use of scleral sutures. It may be replenished over 10 times by puncturing the PDMS layer using 30-gauge needles. The cannula is introduced into the eye by penetrating the sclera. The device's manual mechanical actuation was used to show intraocular medication administration via animal studies conducted on rabbits. Replenish, Inc., based in Pasadena, CA, USA, has created the Micro-Pump System, a technology that enables the precise customization of drug delivery rates. The fundamental configuration remains same with the previous illustration, but with the inclusion of two electrodes inside the replenishable reservoir that establish contact with the medicinal solution (Fig. 3.5). When a pharmaceutical dosage is needed, an electric current is produced between the electrodes, resulting in water electrolysis. As the gases produced by electrolysis are released, the pressure in the reservoir rises, enabling the medication solution to enter the eye via a cannula. The electrolysis process requires minimal energy, making it possible to use small implanted batteries or wireless induction transmission systems instead of traditional power supplies [19]. With MEMS technology, implantable electrodes and iontophoresis, which utilize a direct current power source, can be easily integrated. This method offers the potential to improve medication delivery to the posterior region of the eye. Recent advancements in wearable and flexible electronics have facilitated the development of this technique. Flexible materials and hydrogel materials, commonly used in creating implant devices, can now be used to construct electrodes using the latest technological advancements. Biofuel cells that utilize enzyme-based electrocatalysis at ambient temperature and air pressure can serve as a power source due to their safety and downsizing simplicity. However, leaching and/or enzyme deactivation may shorten the lifetime of biofuel cells that are in a functional state. Drug penetration was enhanced using a transdermal iontophoresis patch for up to six hours when it was powered by an integrated biofuel cell. Biofuel cells may be used as an iontophoresis power source for ophthalmological medication delivery system devices if they can last for many months at least. Electrical stimulation therapy for eye diseases may also make use of implantable electrodes and power sources. While the precise processes of electrical stimulation remain incompletely understood, research has shown the neuroprotective impact of electrical stimulation on retinal cells. In order to enhance the stimulation of retinal cells, it is preferable to locate electrodes in closer proximity to the retina rather than on the surface of the cornea. Combining electrical stimulation with medication delivery in retinal disorders may have a synergistic therapeutic impact, and periocular electrodes would be a preferable choice in terms of electrode design [76, 123, 124].

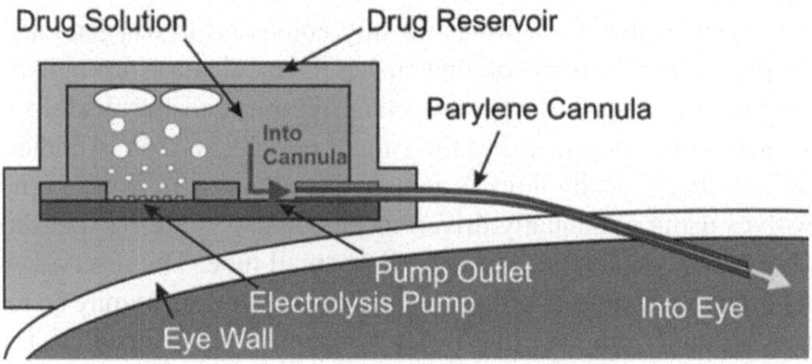

Fig. 3.5 Electrochemical drug delivery pump on a MEMS-based system. Reprinted with permission from Ref. [125]

3.13 Stimuli-Responsive Implants for On-Demand Drug Delivery: BioMEMS Innovations

Stimuli-responsive implants for on-demand drug delivery offer a promising approach to precise and controlled medication administration. These implants are designed to respond to specific external stimuli, such as osmotic pressure, magnetic fields, electrical fields, temperature, or physiological signals, enabling on-demand drug release. In the context of intraocular drug delivery, researchers have explored the use of stimuli-responsive systems to enhance treatment efficacy and minimize side effects (Fig. 3.6). Lo and colleagues introduced the first on-demand microelectromechanical systems (MEMS)-based intraocular medication delivery system. This system included a one-way check valve cannula, adhesive tabs, and a refillable polydimethylsiloxane (PDMS) reservoir. However, the reliability of the refillable PDMS reservoir was found to be limited to 12 refills due to a leak at an injection port. Despite this challenge, the cannula remained unobstructed by biofouling almost six months after implantation. One limitation of the manually operated device was the difficulty in maintaining precise control over the amount of medication injected. To address this issue, hydrolysis was employed to power a small medication pump, enabling more accurate control over the infusion amount. This solution enhanced the feasibility of the intraocular medication delivery system for clinical applications by offering low power consumption and fine control over drug administration. In addition, a commercial mini pump has been developed for clinical use, featuring a refillable tank with a 31-gauge needle, two-way telemetry for wireless control, and a fluid flow sensor. This advanced system offers improved functionality and a maximum lifespan of 5 years [18, 123, 126, 127].

Zhang and colleagues also looked at creating a flexible ocular delivery system that could be inserted behind the eyelid. Simplified surgical procedures, consistent contact, and the ability to target particular locations with a flat electrode are just a few benefits of positioning the device here. It was easy to mass-produce the device, which used electroactive poly(3,4-ethylenedioxythiophene) (PEDOT) electrodes attached to a standard PDMS cup. Thermal effects were evaluated to

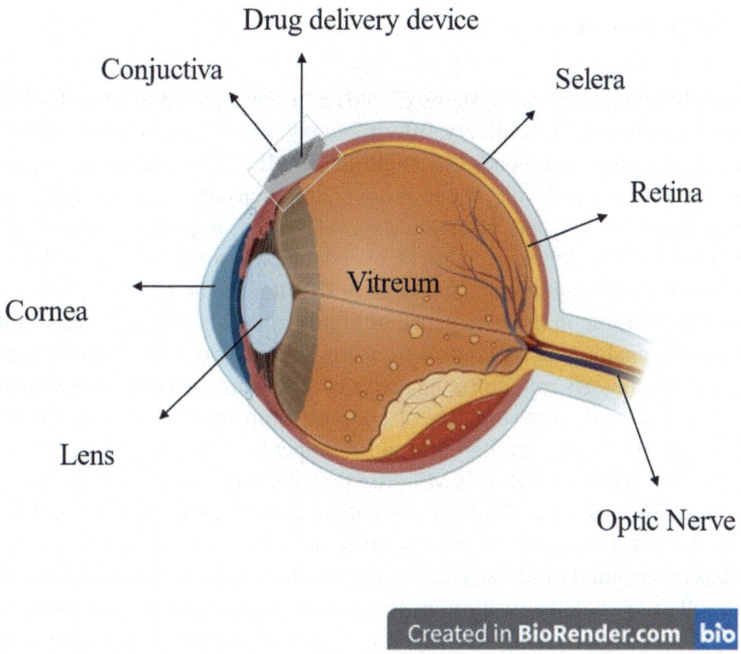

Fig. 3.6 Implantable reservoir and cannula with a check valve made of MEMS technology [125]. Image created with BioRender.com

ensure safety, as iontophoresis, a potential drug delivery method, could increase the temperature around the electrodes. The researchers determined that 37.8 °C was the maximum safe temperature. However, this particular device required the administration of a liquid drug from the outside, and iontophoresis facilitated drug penetration into the vitreous region. Furthermore, the device was not entirely implantable due to the presence of a wire for establishing electrical contact with the PEDOT electrodes [128]. Looking ahead, the integration of stimuli-responsive systems with polymeric implants made of biocompatible materials, such as hydrogels, holds promise for ocular medication delivery. MEMS technology can play a significant role in advancing this field, offering potential solutions to overcome existing challenges and improve treatment outcomes. In summary, stimuli-responsive implants for on-demand drug delivery, particularly in the context of intraocular drug delivery, provide a pathway to precise and controlled medication administration. Ongoing research and development efforts aim to enhance the reliability, functionality, and implantability of these systems, bringing them closer to clinical application and benefiting patients with ocular disorders.

3.14 Conclusion

Bio-microelectromechanical Systems (BioMEMS) have paved the way for remarkable advancements in drug delivery, offering many benefits and addressing the limitations of traditional approaches. The field of BioMEMS has witnessed significant progress, leading to the development of novel drug delivery systems with enhanced efficiency, precision, and patient-centric approaches. The importance and advantages of BioMEMS in drug delivery cannot be overstated. Microfabrication methods have been adapted to create materials and devices that mimic the physiological scale of cells in vivo, resulting in the emergence of highly innovative drug delivery systems. These systems enable precise control over drug release, complex delivery patterns, and accurate dosing. By miniaturizing active components such as valves and pumps, BioMEMS enables the construction of compact, implantable, or portable devices that offer improved drug availability and therapeutic outcomes. Throughout this chapter, various aspects of BioMEMS drug delivery have been explored. Oral drug delivery systems have been developed to enhance medication retention and interaction with the intestinal wall, improving absorption and efficacy. Micro-patch technology has contributed to advancements in fabrication techniques, enabling efficient and controlled drug delivery through innovative patch designs. Bio-adhesion has played a crucial role in drug delivery, leveraging BioMEMS to enhance adhesion properties and improve the targeting of specific tissues or organs. Wireless endoscopic capsules have revolutionized drug delivery monitoring by providing real-time insights into drug distribution and pharmacokinetics, allowing for precise adjustments and personalized treatment regimens. External systems, such as microneedles, have introduced non-invasive drug delivery methods that offer painless administration and targeted drug delivery. Microneedle arrays have expanded the possibilities of transdermal drug delivery, allowing for the delivery of macromolecules and vaccines with improved efficiency and patient comfort. Implantable drug delivery systems have also benefited greatly from BioMEMS technology. These systems offer long-term, controlled drug release and can be tailored to specific patient needs. Ongoing research focuses on improving patient compliance, device size, drug delivery rates, and biocompatibility. Integration of sensor technology in these devices enables real-time monitoring of physiological changes, ensuring precise and personalized medication administration. Although challenges remain, including scaling-up processes, ensuring long-term performance stability, and careful material selection, the future of drug delivery lies in the continued advancement of microfabrication techniques and BioMEMS. These innovations hold immense potential to transform the landscape of personalized medicine, leading to improved patient outcomes, enhanced therapeutic efficacy, and a higher quality of life. In conclusion, the field of BioMEMS is on the brink of revolutionizing drug delivery by offering innovative solutions that overcome the limitations of traditional approaches. The integration of microfabrication techniques, advanced materials, and sensor technology has the potential to redefine the way medications are administered, monitored, and tailored to individual patients. As research progresses, in vivo testing and meticulous consideration of biocompatibility will be crucial in translating these advancements into clinical applications. Ultimately, BioMEMS-driven drug delivery systems have the power to reshape healthcare by enabling personalized, precise, and effective therapies that improve the lives of patients worldwide.

References

1. Goel, V.K., Ferrara, L.: Basic science symposium II: MEMS technology. SAS J. **2**(2), 120–129 (2008)
2. Garcia-Ramirez, R., Hosseini, S.: History of bio-microelectromechanical Systems (BioMEMS). In: Hosseini, S., et al. (eds.) BioMEMS: Biosensing Applications, pp. 1–20. Springer Singapore, Singapore (2021)
3. Saliterman, S.: Fundamentals of BioMEMS and Medical Microdevices, vol. 153. SPIE press (2006)
4. Folch, A.: Introduction to bioMEMS. CRC press (2016)
5. Cerda-Kipper, A.S., Hosseini, S.: Bio-microelectromechanical Systems (BioMEMS) in Biosensing Applications—Chemiluminescence Detection Strategies, pp. 123–151. BioMEMS (2020)
6. Cros, F.: BioMEMS. In: Narayan, R. (ed.) Biomedical Materials, pp. 581–620. Springer International Publishing, Cham (2021)
7. Borenstein, J.T., et al.: Bionanotechnology and bioMEMS (BNM): state-of-the-art applications, opportunities, and challenges. Lab Chip **23**(23), 4928–4949 (2023)
8. Singh, A.V., et al.: Emerging technologies for in vitro inhalation toxicology. Adv. Healthcare Mater. **10**(18), 2100633 (2021)
9. Kulkarni, M.B., Goel, S.: Recent advancements in integrated microthermofluidic systems for biochemical and biomedical applications—a review. Sens. Actuators, A **341**, 113590 (2022)
10. Farahani, A., et al.: Silk-based biopolymers promise extensive biomedical applications in tissue engineering, drug delivery, and BioMEMS. J. Polym. Environ. **31**(11), 4559–4582 (2023)
11. Rahmanian, M., et al.: A micropillar array-based microfluidic chip for label-free separation of circulating tumor cells: the best micropillar geometry? J. Adv. Res. **47**, 105–121 (2023)
12. Sartipzadeh, O., et al.: Microfluidic-assisted synthesis and modeling of stimuli-responsive monodispersed chitosan microgels for drug delivery applications. Sci. Rep. **12**(1), 8382 (2022)
13. Sartipzadeh, O., et al.: Controllable size and form of droplets in microfluidic-assisted devices: effects of channel geometry and fluid velocity on droplet size. Mater. Sci. Eng. C **109**, 110606 (2020)
14. Sartipzadeh, O., et al.: Microfluidic-assisted synthesis and modelling of monodispersed magnetic nanocomposites for biomedical applications. **9**(1), 1397–1407 (2020)
15. Pandey, Y., Singh, S.P.: Recent advances in Bio-MEMS and future possibilities: an overview. J Inst Eng (India): Ser B **104**(6), 1377–1388 (2023)
16. Liu, X., et al.: Smart contact lenses for healthcare monitoring and therapy. ACS Nano **18**(9), 6817–6844 (2024)
17. Enciu, C.-C., et al.: Application of micro electrical discharge machining and electrochemical machining in manufacturing of micro-electromechanical systems: a review. Macromol. Symp. **404**(1), 2100449 (2022)
18. Bhavana, H.T., et al.: Recent advances in drug delivery systems: MEMS perspective. Disruptive Developments in Biomedical Applications, pp. 299–309 (2022)
19. Basu, A.K., et al.: MEMS Applications in Biology and Healthcare. AIP Publishing Melville (2021)

20. Barua, R., et al.: Chapter 14—advances in MEMS micropumps and their emerging drug delivery and biomedical applications. In: Nayak, A.K., et al. (eds.) Advances and Challenges in Pharmaceutical Technology, pp 411–452. Academic Press (2021)
21. Song, H., et al.: Miniature FBG vibration sensor with high performance and low angle dependence for two-dimensional vibration measurement. IEEE Trans. Instrum. Meas. **72**, 1–11 (2023)
22. Li, J., et al.: Three-dimensional vascular reconstruction and Doppler flow measurement using PMUTs. In: 2022 IEEE International Ultrasonics Symposium (IUS) (2022)
23. Yu, J., et al.: Mechanically sensing and tailoring electronic properties in two-dimensional atomic membranes. Curr. Opin. Solid State Mater. Sci. **25**(2), 100900 (2021)
24. Xuan, X., Qian, S.: Micro/Nano-Chip Electrokinetics, vol III. MDPI (2021)
25. Huang, T.-Y., Gu, H., Nelson, B.J.: Increasingly intelligent micromachines. Annu. Rev. Control, Rob., Auton. Syst. **5**, 279–310 (2022)
26. Nuxoll, E.E., Siegel, R.A.: BioMEMS devices for drug delivery. IEEE Eng. Med. Biol. Mag. **28**(1), 31–39 (2009)
27. Hua, S.: Advances in oral drug delivery for regional targeting in the gastrointestinal tract—influence of physiological, pathophysiological and pharmaceutical factors. Front Pharmacol **11** (2020)
28. Vinarov, Z., et al.: Impact of gastrointestinal tract variability on oral drug absorption and pharmacokinetics: an UNGAP review. Eur. J. Pharm. Sci. **162**, 105812 (2021)
29. Stillhart, C., et al.: Impact of gastrointestinal physiology on drug absorption in special populations—an UNGAP review. Eur. J. Pharm. Sci. **147**, 105280 (2020)
30. Xiong, K., et al.: Construction of food-grade pH-sensitive nanoparticles for delivering functional food ingredients. Trends Food Sci. Technol. **96**, 102–113 (2020)
31. Gaohua, L., Miao, X., Dou, L.: Crosstalk of physiological pH and chemical pKa under the umbrella of physiologically based pharmacokinetic modeling of drug absorption, distribution, metabolism, excretion, and toxicity. Expert Opin. Drug Metab. Toxicol. **17**(9), 1103–1124 (2021)
32. Shahidi, M., et al.: HB5 aptamer-tagged graphene oxide for co-delivery of doxorubicin and silibinin, and highly effective combination therapy in breast cancer. Cancer Nanotechnol. **14**(1), 59 (2023)
33. Yaghoubi, F., et al.: A functionalized graphene oxide with improved cytocompatibility for stimuli-responsive co-delivery of curcumin and doxorubicin in cancer treatment. Sci. Rep. **12**(1), 1959 (2022)
34. Yaghoubi, F., et al.: Multiresponsive carboxylated graphene oxide-grafted aptamer as a multifunctional nanocarrier for targeted delivery of chemotherapeutics and bioactive compounds in cancer therapy. Nanotechnol. Rev. **10**(1), 1838–1852 (2021)
35. Mazidi, Z., et al.: Smart stimuli-responsive implantable drug delivery systems for programmed and on-demand cancer treatment: an overview on the emerging materials. Chem. Eng. J. **433**, 134569 (2022)
36. Abtahi, N.A., et al.: Development of highly efficient niosomal systems for co-delivery of drugs and genes to treat breast cancer in vitro and in vivo. Cancer Nanotechnol. **13**(1), 1–18 (2022)
37. Abtahi, N.A., et al.: A comparative study on biopharmaceutical function of curcumin and miR-34a by multistimuli-responsive nanoniosome carrier: in-vitro and in-vivo. Front. Mol. Biosci. **9**, 1043277 (2022)
38. Abtahi, N.A., et al.: Smart stimuli-responsive biofunctionalized niosomal nanocarriers for programmed release of bioactive compounds into cancer cells in vitro and in vivo. Nanotechnol. Rev. **10**(1), 1895–1911 (2021)
39. Abtahi, N.A., et al.: Multi-sensitive functionalized niosomal nanocarriers for controllable gene delivery in vitro and in vivo. Cancer Nanotechnol. **14**(1), 1–18 (2023)
40. Bhutani, U., Basu, T., Majumdar, S.: Oral drug delivery: conventional to long acting new-age designs. Eur. J. Pharm. Biopharm. **162**, 23–42 (2021)

41. Brayden, D.J., et al.: Systemic delivery of peptides by the oral route: formulation and medicinal chemistry approaches. Adv. Drug Deliv. Rev. **157**, 2–36 (2020)
42. Gupta, J., Gupta, R., Vanshita.: Microneedle technology: an insight into recent advancements and future trends in drug and vaccine delivery. ASSAY Drug Dev. Technol. **19**(2), 97–114 (2020)
43. Kulkarni, D., et al.: Recent advancements in microneedle technology for multifaceted biomedical applications. Pharmaceutics **14** (2022). https://doi.org/10.3390/pharmaceutics14051097
44. Nagarkar, R., et al.: A review of recent advances in microneedle technology for transdermal drug delivery. J. Drug Delivery Sci. Technol. **59**, 101923 (2020)
45. Milián-Guimerá, C., et al.: Smart pills and drug delivery devices enabling next generation oral dosage forms. J. Control. Release **364**, 227–245 (2023)
46. Bruschi, M.L., de Souza Ferreira, S.B., Bassi da Silva, J.: Chapter 4—Mucoadhesive and mucus-penetrating polymers for drug delivery. In: Martins, J.P., Santos, H.A., (eds.) Nanotechnology for Oral Drug Delivery, pp. 77–141. Academic Press (2020)
47. Alagusundaram, M., et al.: Development and characterization of gel-based buccoadhesive bilayer formulation of nifedipine. Gels **9** (2023). https://doi.org/10.3390/gels9090688
48. Dathathri, E., et al.: Fabrication of low-cost composite polymer-based micro needle patch for transdermal drug delivery. Appl. Nanosci. **10**(2), 371–377 (2020)
49. Suresh Chinnathampy, M., Aruna, T., Muthukumaran, N.: Design and fabrication of micro strip patch antenna for cognitive radio applications. Wireless Pers. Commun. **121**(3), 1577–1592 (2021)
50. Chang, K.-T., et al.: Optimal design and fabrication of a microneedle arrays patch. J. Manuf. Process. **54**, 274–285 (2020)
51. Sonetha, V., Majumdar, S., Shah, S.: Step-wise micro-fabrication techniques of microneedle arrays with applications in transdermal drug delivery—a review. J. Drug Deliv. Sci. Technol. **68**, 103119 (2022)
52. Oh, N.G., Hwang, S.Y., Na, Y.H.: Fabrication of a PVA-based hydrogel microneedle patch. ACS Omega **7**(29), 25179–25185 (2022)
53. Jamaledin, R., et al.: Advances in antimicrobial microneedle patches for combating infections. Adv. Mater. **32**(33), 2002129 (2020)
54. Wang, Y., et al.: Emerging functional biomaterials as medical patches. ACS Nano **15**(4), 5977–6007 (2021)
55. Qiu, S., et al.: Recent advances in surface manipulation using micro-contact printing for biomedical applications. Smart Mater. Med. **2**, 65–73 (2021)
56. Mishra, R., Bhattacharyya, T.K.: MEMS-based Transdermal Drug Delivery. CRC Press (2023)
57. Park, S., et al.: Micro-groove patterned PCL patches with DOPA for rat Achilles tendon regeneration. J. Ind. Eng. Chem. **105**, 352–364 (2022)
58. Dastidar, M.G., Roy, S., Neogi, S.: Bioadhesives for the advancement of controlled drug delivery and wearable bioelectronics. In: Biobased Adhesives, pp. 705–725 (2023)
59. Singh, I., et al.: Bioadhesive films as drug delivery systems. Drug Deliv. Lett. **11**(1), 2–15 (2021)
60. Park, J., et al.: Conformal fixation strategies and bioadhesives for soft bioelectronics. Adv. Funct. Mater. **n/a**(n/a), 2313728 (2024)
61. Habib, R., et al.: Thiolated chitosan microneedle patch of levosulpiride from fabrication, characterization to bioavailability enhancement approach. Polymers, **14** (2022). https://doi.org/10.3390/polym14030415
62. Li, X.-J., et al.: Composite dissolvable microneedle patch for therapy of oral mucosal diseases. Biomater. Adv. **139**, 213001 (2022)
63. Ahmad, Z., et al.: Fabrication and characterization of thiolated chitosan microneedle patch for transdermal delivery of tacrolimus. AAPS PharmSciTech **21**(2), 68 (2020)
64. Nizar Hammed, M.A., Fayadh, R.A., Farhan, H.M.: UWB pentagonal shaped fractal patch antenna for wireless capsule endoscopy. IOP Conf. Ser.: Mater. Sci. Eng. **745**(1), 012092 (2020)

65. Cortegoso Valdivia, P., et al.: An overview of robotic capsules for drug delivery to the gastrointestinal tract. J. Clin. Med. **10** (2021). https://doi.org/10.3390/jcm10245791
66. McGirr, M.E.A., et al.: The use of the intelisite® companion device to deliver mucoadhesive polymers to the dog colon. Eur. J. Pharm. Sci. **36**(4), 386–391 (2009)
67. Byrne, J., et al.: Devices for drug delivery in the gastrointestinal tract: a review of systems physically interacting with the mucosa for enhanced delivery. Adv. Drug Deliv. Rev. **177**, 113926 (2021)
68. Rehan, M., et al.: Anchoring mechanism for capsule endoscope: mechanical design, fabrication and experimental evaluation. Micromachines **13** (2022). https://doi.org/10.3390/mi13122045
69. Zingales, S.K., Ferguson, J.H.: Chapter 9—AI in microfabrication technology. In: Philip, A., et al. (eds.) A Handbook of Artificial Intelligence in Drug Delivery, pp. 213–239. Academic Press (2023)
70. Rodrigue, H., Kim, J.: Soft actuators in surgical robotics: a state-of-the-art review. Intel. Serv. Robot. **17**(1), 3–17 (2024)
71. Fakhr Abdollahi, A., et al.: A review of modeling and control of remote-controlled capsule endoscopes. Expert Rev. Med. Devices **21**(4), 293–306 (2024)
72. Uboldi, M., et al.: Administration strategies and smart devices for drug release in specific sites of the upper GI tract. J. Control. Release **348**, 537–552 (2022)
73. Lyutakov, I., Penchev, P.: Current advances in drug delivery systems for capsule endoscopy. Curr. Drug Metab. **21**(11), 838–843 (2020)
74. Emami, A., Sharma, S.: 3D localization of miniaturized devices for precision medicine: enabling wireless surgical navigation, health monitoring, and targeted drug delivery. IEEE Solid-State Circuits Mag. **15**(2), 41–48 (2023)
75. Liao, Z., Zoumhani, O., Boutry, C.M.: Recent advances in magnetic polymer composites for BioMEMS: a review. Materials **16** (2023). https://doi.org/10.3390/ma16103802
76. Zhang, H., Jackson, J.K., Chiao, M.: Microfabricated drug delivery devices: design, fabrication, and applications. Adv. Func. Mater. **27**(45), 1703606 (2017)
77. Qureshi, W.A.J.N.R.D.D.: Current and future applications of the capsule camera. **3**(5), 447–450 (2004)
78. Wang, R., et al.: Recent advances in polymer microneedles for drug transdermal delivery: design strategies and applications. Macromol. Rapid Commun. **43**(8), 2200037 (2022)
79. Lee, K., et al.: Non-transdermal microneedles for advanced drug delivery. Adv. Drug Deliv. Rev. **165–166**, 41–59 (2020)
80. Gowda, B.H.J., Ahmed, M.G., Sanjana, A.: Can microneedles replace hypodermic needles? Resonance **27**(1), 63–85 (2022)
81. Donnelly, R.F., Prausnitz, M.R.: The promise of microneedle technologies for drug delivery. Drug Deliv. Transl. Res. **14**(3), 573–580 (2024)
82. Avcil, M., Çelik, A.: Microneedles in drug delivery: progress and challenges. Micromachines **12** (2021). https://doi.org/10.3390/mi12111321
83. Queiroz, M.L.B., et al.: Microneedles as an alternative technology for transdermal drug delivery systems: a patent review. Expert Opin. Ther. Pat. **30**(6), 433–452 (2020)
84. Sabbagh, F., Kim, B.S.: Recent advances in polymeric transdermal drug delivery systems. J. Control. Release **341**, 132–146 (2022)
85. Jeong, W.Y., et al.: Recent advances in transdermal drug delivery systems: a review. Biomater. Res. **25**(1), 24
86. Ng, L.C., Gupta, M.: Transdermal drug delivery systems in diabetes management: a review. Asian J. Pharm. Sci. **15**(1), 13–25 (2020)
87. Akhtar, N., et al.: Non-invasive drug delivery technology: development and current status of transdermal drug delivery devices, techniques and biomedical applications. **65**(3), 243–272 (2020)
88. Ramadon, D., et al.: Enhancement strategies for transdermal drug delivery systems: current trends and applications. Drug Deliv. Transl. Res. **12**(4), 758–791 (2022)

89. Azmana, M., et al.: Transdermal drug delivery system through polymeric microneedle: a recent update. J. Drug Delivery Sci. Technol. **60**, 101877 (2020)
90. Jung, J.H., Jin, S.G.: Microneedle for transdermal drug delivery: current trends and fabrication. J. Pharm. Investig. **51**(5), 503–517 (2021)
91. Tariq, N., Ashraf, M.W., Tayyaba, S.: A review on solid microneedles for biomedical applications. J. Pharm. Innov. **17**(4), 1464–1483 (2022)
92. Economidou, S.N., et al.: Optimisation of design and manufacturing parameters of 3D printed solid microneedles for improved strength, sharpness, and drug delivery. Micromachines **12** (2021). https://doi.org/10.3390/mi12020117
93. Manoj, H., et al.: Microneedles: current trends and applications. In: Microfluidics and bio-MEMS, pp. 275–342. Jenny Stanford Publishing (2020)
94. Babu, M.R., et al.: Unravelling the role of microneedles in drug delivery: principle, perspectives, and practices. Drug Deliv. Transl. Res. **14**(6), 1393–1431 (2024)
95. Meng, F., et al.: Polymeric-based microneedle arrays as potential platforms in the development of drugs delivery systems. J. Adv. Res. **26**, 137–147 (2020)
96. Dabholkar, N., et al.: Biodegradable microneedles fabricated with carbohydrates and proteins: revolutionary approach for transdermal drug delivery. Int. J. Biol. Macromol. **170**, 602–621 (2021)
97. Tucak, A., et al.: Microneedles: characteristics, materials, production methods and commercial development. Micromachines **11** (2020). https://doi.org/10.3390/mi11110961
98. Aldawood, F.K., Andar, A., Desai, S.: A comprehensive review of microneedles: types, materials, processes, characterizations and applications. Polymers **13** (2021). https://doi.org/10.3390/polym13162815
99. Mdanda, S., et al.: Recent advances in microneedle platforms for transdermal drug delivery technologies. Polymers **13** (2021). https://doi.org/10.3390/polym13152405
100. Sirbubalo, M., et al.: 3D printing—a touch-button approach to manufacture microneedles for transdermal drug delivery. Pharmaceutics **13** (2021). https://doi.org/10.3390/pharmaceutics13070924
101. Yang, D., et al.: Microneedle-mediated transdermal drug delivery for treating diverse skin diseases. Acta Biomater. **121**, 119–133 (2021)
102. Salwa, et al.: Polymeric microneedles for transdermal delivery of nanoparticles: frontiers of formulation, sterility and stability aspects. J. Drug Delivery Sci. Technol. **65**, 102711 (2021)
103. Chakraborty, C., Bhattacharya, M., Lee, S.-S.: Current status of microneedle array technology for therapeutic delivery: from bench to clinic. Mol. Biotechnol. (2023)
104. Nguyen, H.X., Nguyen, C.N.: Microneedle-mediated transdermal delivery of biopharmaceuticals. Pharmaceutics **15** (2023). https://doi.org/10.3390/pharmaceutics15010277
105. Athirathinam, K., Nandakumar, S., Kandasamy, R.: Biopolymers and osmolytes—a focus towards the prospects of stability and adjuvanticity of vaccines. Macromol. Res. **30**(9), 599–608 (2022)
106. Yang, L., et al.: Polymeric microneedle-mediated sustained release systems: design strategies and promising applications for drug delivery. Asian J. Pharm. Sci. **17**(1), 70–86 (2022)
107. Sartawi, Z., Blackshields, C., Faisal, W.: Dissolving microneedles: applications and growing therapeutic potential. J. Control. Release **348**, 186–205 (2022)
108. Bauleth-Ramos, T., et al.: Recent approaches for enhancing the performance of dissolving microneedles in drug delivery applications. Mater. Today **63**, 239–287 (2023)
109. Moore, L.E., Vucen, S., Moore, A.C.: Trends in drug- and vaccine-based dissolvable microneedle materials and methods of fabrication. Eur. J. Pharm. Biopharm. **173**, 54–72 (2022)
110. Kenchegowda, M., et al.: Tiny titans- unravelling the potential of polysaccharides and proteins based dissolving microneedles in drug delivery and theranostics: a comprehensive review. Int. J. Biol. Macromol. **253**, 127172 (2023)
111. Villarruel Mendoza, L.A., et al.: Recent advances in micro-electro-mechanical devices for controlled drug release applications. Front. Bioeng. Biotechnol. **8** (2020)

112. Yang, W., Gong, Y., Li, W.: A review: electrode and packaging materials for neurophysiology recording implants. Front. Bioeng. Biotechnol. **8** (2021)
113. Gupta, H.S., Sethi, S.K., Verma, A.: Applications of coating materials: a critical overview. Coating Materials: Computational Aspects, Applications and Challenges, pp. 81–109 (2023)
114. Prestat, M., Thierry, D.: Corrosion of titanium under simulated inflammation conditions: clinical context and in vitro investigations. Acta Biomater. **136**, 72–87 (2021)
115. Rav Acha, M., Soifer, E., Hasin, T.: Cardiac implantable electronic miniaturized and micro devices. Micromachines **11** (2020). https://doi.org/10.3390/mi11100902
116. Magill, E., et al.: Solid implantable devices for sustained drug delivery. Adv. Drug Deliv. Rev. **199**, 114950 (2023)
117. Fayzullin, A., et al.: Implantable drug delivery systems and foreign body reaction: traversing the current clinical landscape. Bioengineering **8** (2021). https://doi.org/10.3390/bioengineering8120205
118. Zhang, J., et al.: Wearable glucose monitoring and implantable drug delivery systems for diabetes management. Adv. Healthcare Mater. **10**(17), 2100194 (2021)
119. Kumar, A., Pillai, J.: Chapter 13—implantable drug delivery systems: an overview. In: Grumezescu, A.M., (ed.) Nanostructures for the Engineering of Cells, Tissues and Organs, pp. 473–511. William Andrew Publishing (2018)
120. Jain, K.K.: Drug delivery systems-an overview. Drug Delivery Systems, pp. 1–50 (2008)
121. Stewart, S.A., et al.: Implantable polymeric drug delivery devices: classification, manufacture, materials, and clinical applications. Polymers **10** (2018). https://doi.org/10.3390/polym10121379
122. Talebian, S., et al.: Biopolymers for antitumor implantable drug delivery systems: recent advances and future outlook. Adv. Mater. **30**(31), 1706665 (2018)
123. Lee, H.J., et al.: MEMS devices for drug delivery. Adv. Drug Deliv. Rev. **128**, 132–147 (2018)
124. Khandan, O.: Titanium MEMS Technology Development for Drug Delivery and Microfluidic Applications. University of California, Riverside (2015)
125. Li, P.-Y., et al.: An electrochemical intraocular drug delivery device. **143**(1), 41–48 (2008)
126. Caldorera-Moore, M., Peppas, N.A.: Micro- and nanotechnologies for intelligent and responsive biomaterial-based medical systems. Adv. Drug Deliv. Rev.Deliv. Rev. **61**(15), 1391–1401 (2009)
127. James, J.J., Sandhya, K.V.: Integration of biosensors and drug delivery systems for biomedical applications. In: Applied Artificial Intelligence, pp. 353–372. CRC Press (2023)
128. Zhang, P., et al.: Electrochemical and electrical biosensors for wearable and implantable electronics based on conducting polymers and carbon-based materials. Chem. Rev. **124**(3), 722–767 (2024)

Chapter 4
Microfluidic Micropumps: Advancing Biomedical Applications with Micro Electromechanical-Based Systems (MEMS)

Abstract Microfluidic micropumps utilizing Micro Electromechanical Systems (MEMS) technology have emerged as a promising tool for advancing biomedical applications, particularly in medication delivery and actuation purposes. This chapter provides a brief overview of the different types of micropumps and their components, as well as their performance characteristics. Micropumps can be broadly classified into two categories based on their actuation mechanisms: mechanical and non-mechanical pumps. Mechanical micropumps rely on physical displacement to generate fluid flow, while non-mechanical micropumps utilize alternative energy sources for actuation. The acquisition of actuation energy is a critical factor in driving fluid flow within these micropumps. In this chapter, we aim to introduce the concept of micro-pumps and provide insights into their components and performance attributes. The potential applications of microfluidic micropumps in medication delivery and various biomedical fields are also discussed. By leveraging MEMS technology, these micropumps offer precise control, miniaturization capabilities, and enhanced functionality, enabling advancements in targeted drug delivery, diagnostics, and other biomedical applications. Understanding the design principles, operational mechanisms, and performance characteristics of microfluidic micropumps is essential for harnessing their full potential in biomedical research and healthcare. Further research and development efforts are required to optimize their performance, explore new applications, and address challenges associated with integration, reliability, and compatibility with biological systems. By leveraging the advantages of MEMS-based micropumps, we can unlock new possibilities for improved therapeutic outcomes and advancements in biomedical sciences.

Keywords Microfluidics · Micropumps · Biomedical applications · MEMS

S. M. Naghib and Y. Rezaian, *Frontiers of Nanobiotechnology*, Series in BioEngineering, https://doi.org/10.1007/978-3-031-88419-1_4

4.1 Introduction

Microfluidic systems include the manipulation and regulation of tiny volumes of fluid flow. A notable branch within microfluidics is the development of microfluidic devices based on Micro Electromechanical Systems (MEMS), which have garnered considerable interest in the biomedical domain. The objective of this chapter is to present a thorough examination of microfluidic devices, with a particular emphasis on micropumps, in the context of biological applications. The chapter covers various aspects related to micropumps, including their material composition, construction, testing, safety considerations, applications, failure analysis, performance parameters, actuation strategies, working principles, monetization concerns, and prospects for the future [1, 2]. The term "micropump" denotes a miniaturized gas or liquid pumping apparatus capable of achieving flow rates below a milliliter per minute [3]. Micropumps play a crucial role in providing the necessary energy for fluid movement within microfluidic devices. The initial studies on micropumps for medication delivery date back to the mid-1970s when Spencer and his colleagues developed an improved version of Thomas and Bessman's original design [3, 4]. Since then, micropumps have undergone significant advancements and now offer several advantages, including compact size, lightweight, high mobility, a wide range of flow rates, low cost, low power consumption, and compatibility with integration into other microfluidic devices [5].

The rapid growth of engineering in the sub-millimeter and small-scale domain has led to the development of miniaturized devices and systems. The primary goal of miniaturization is to reduce the size, cost, and improve the performance of a system. Downsized devices bring numerous advantages, such as low energy requirements, reduced sample volume, the use of disposable materials, portability, affordability, and increased speed. Researchers have recognized the potential benefits of miniaturization, leading to the development of small systems capable of managing liquids and fluids on the microscale to the nanoscale. Microfluidics emerged as a field that combines micro-scale engineering and fluid mechanics to manipulate and analyze minute amounts of fluids. Microfluidic devices act as remote pumping systems, enabling monitoring, pumping, mixing, and regulation of tiny fluid volumes [6]. It is important to note that fluid behavior at the micro and nanoscale differs significantly from that at the macro scale. Surface tension plays a crucial role in microfluidic systems, and conventional fluidic systems may not provide a realistic representation of sample flow when the size of biological samples becomes comparable to the diameters of flow channels or needles. The development of microfluidic components, devices, methods, and manufacturing processes has so attracted a great deal of interest in recent years. The advancement of MEMS and NEMS technologies has made it possible to manufacture high-performance, small-sized medical tools, addressing critical medical needs such as enhanced bioavailability, controlled administration with minimal side effects, and improved therapeutic efficacy [7].

Micropumps play a vital role as key components in microfluidic systems, aiming to deliver precise doses of medication at specific rates and durations to targeted tissues. Micropumps are responsible for generating fluid motion within the system [8]. Compact and small micropumps can be integrated into drug delivery systems to precisely regulate the dose administered to the body. For instance, micropumps can be programmed to deliver medications consistently throughout the day, preventing any fluctuations or gaps in the drug concentration in the patient's bloodstream. Achieving accurate dosage and timing is crucial for studying the effects of medication on the body, highlighting the importance of selecting the right micropump for optimal device performance [9]. A drug delivery system (DDS) consists of various components, including a drug reservoir, microchannels, microsensors, valves, micropumps, and associated circuits [10–14]. Micropumps serve as essential components within the DDS, ensuring accurate and efficient transfer of the drug from the reservoir to the target location within the body, such as a blood artery or tissue [15]. Although oral medicines, inhalers, and subcutaneous injections are standard drug delivery techniques, they may not consistently and effectively deliver all pharmaceuticals to their intended therapeutic areas. This can lead to initial high drug concentrations in the bloodstream, exceeding the toxicity limit, followed by a gradual decline to an ineffective level over time, necessitating frequent medication intake [16]. Controlled drug delivery systems offer a solution to this challenge [17–25]. By utilizing a microactuator device, such as a micropump, in conjunction with a controller, the appropriate and effective amount of medication can be accurately computed and delivered at the precise moment. Benefits of controlled medication release include improved treatment effectiveness, fewer side effects, and site-specific drug delivery [15]. Moreover, an implanted DDS may reduce the danger of contamination or illness linked to conventional injection techniques and ease the pain associated with frequent injections. Micropumps play a crucial role in DDS, and various mechanical micropumps with diverse actuation principles, including thermo-pneumatic, electrostatic, shape memory alloy (SMA), electromagnetic, and piezoelectric, have been developed over the years. The advantages of piezoelectric actuation, such as moderate pressure and displacement with low power consumption, high reliability, and energy efficiency, make them particularly desirable for medical applications [16, 26].

Based on their actuation mechanisms, micropumps can be divided into two categories: non-mechanical micropumps and mechanical micropumps [7]. Non-mechanical micropumps include piezoelectric micropumps, thermal actuation micropumps, electrostatic micropumps, shape memory alloy micropumps, ion conductive polymer film micropumps, electromagnetic micropumps, and bimetallic micropumps [5, 6, 27, 28]. These micropumps operate based on principles such as electro-wetting, evaporation, bubble formation, electroosmosis, electrohydrodynamics, and magnetohydrodynamics. On the other hand, mechanical micropumps rely on mechanical components for fluid actuation.

In summary, microfluidic systems and microfluidic devices, particularly micropumps, have gained significant attention in biomedical applications. Micropumps play a crucial role in delivering precise doses of medication and controlling fluid flow within microfluidic systems. The advancements in MEMS and NEMS technologies have paved the way for miniaturized devices that offer various advantages such as improved performance, reduced cost, and enhanced portability. Micropumps can be categorized based on their actuation mechanisms, and both non-mechanical and mechanical micropumps have been developed to cater to different application requirements. The integration of micropumps into drug delivery systems enables controlled and targeted drug release, resulting in improved therapeutic efficacy and reduced side effects.

4.2 Explanation of Microelectromechanical Systems (MEMS) and Their Role in Micropump Design

The field of Microelectromechanical systems (MEMS) has completely transformed the way micropumps are designed by integrating tiny mechanical and electrical components onto a single chip. This interdisciplinary domain combines the art of microfabrication, microelectronics, and mechanical engineering principles to create devices that can be used in a variety of applications. Specifically in the realm of micropump design, MEMS technology plays a crucial role in achieving miniaturization and seamlessly integrating fluidic systems. MEMS-based micropumps are meticulously engineered to manipulate and control the flow of fluids on a remarkably small scale, often dealing with volumes as minuscule as microliters or nanoliters. In comparison to their larger counterparts, these devices offer a multitude of benefits, including enhanced control, reduced sample and reagent volumes, faster response times, and compatibility with lab-on-a-chip systems. By integrating sensors and control electronics, the functionality and performance of MEMS micropumps are further improved, enabling real-time monitoring and feedback control. On the whole, MEMS technology has opened up a whole new realm of possibilities for highly efficient, compact, and precise micropumps, thereby contributing to advancements in biomedical applications, drug delivery systems, diagnostics, and microfluidic devices [29, 30].

4.2.1 Design and Fabrication of MEMS Micropumps

The manufacturing methods of MEMS micropumps are similar to those of microelectromechanical systems discussed in previous chapters. These fabrication techniques, such as lithography, etching, and bonding, are commonly used in the production of MEMS-based micropumps. They allow for the creation of intricate structures, precise fluid channels, and reliable bonding between different components, ultimately enhancing the functionality and performance of MEMS micropumps [15].

4.2.2 Design Considerations for MEMS Micropumps in Biomedical Applications

Micropumps based on MEMS technology for medical applications are constructed utilizing a wide range of materials. The utilization of silicon is prevalent due to its exceptional mechanical properties, compatibility with microfabrication processes, and biocompatibility. Glass is the preferred choice owing to its transparency, biocompatibility, and chemical resistance, rendering it suitable for the creation of microfluidic channels and chambers. Polymers, such as PDMS, offer characteristics such as flexibility, biocompatibility, and ease of fabrication. Metals, such as gold, platinum, titanium, and aluminum, are employed for the production of electrodes, valves, and actuators. Ceramics, like alumina and zirconia, can be utilized for their high-temperature resistance or biocompatibility [31].

When designing micropumps, it is crucial to consider [32, 33]:

Fluid compatibility to prevent contamination or degradation.
Biocompatibility to ensure safety in biological systems.
Miniaturization for easy implantation and portability.
Precise flow control through valves and sensors.
Low power consumption for extended use.
Reliability and longevity with robust designs and fault detection mechanisms.
Serializability to maintain aseptic conditions.
Integration with sensors and control systems for closed-loop control and monitoring.
Manufacturability and scalability for mass production.
Regulatory compliance to meet safety and quality standards.

4.3 Types of MEMS Micropumps

Micropumps can be broadly categorized into two primary classifications: mechanical and non-mechanical. Each classification possesses its distinctive operational principles and applications.

1. Mechanical Micropumps: Mechanical micropumps employ mechanical constituents such as valves, membranes, or actuators to generate fluid flow. They depend on physical displacement or pumping mechanisms to induce fluid motion.
2. Non-Mechanical Micropumps: Non-mechanical micropumps do not rely on mechanical constituents for fluid propulsion. Instead, they harness various physical or electrokinetic phenomena to propel the fluid [6].

4.3.1 Mechanical Micropumps

Mechanical micropumps operate by utilizing various mechanical components, such as valves, membranes, or actuators, to generate fluid movement. They rely on physical displacement or pumping mechanisms to efficiently transport fluids. Within the field of mechanical micropumps, there are several commonly used types, each with its distinct characteristics and applications [5, 34].

1. Diaphragm Micropumps: One prevalent type is the diaphragm micropump, which utilizes a flexible diaphragm that is actuated to create pressure variations, thereby propelling the fluid. These micropumps can be driven by different mechanisms, including piezoelectric, electrostatic, or electromagnetic actuation, providing versatility in their operation.
2. Peristaltic Micropumps: Another notable variety is the peristaltic micropump, which employs a series of rollers or compressing elements to squeeze a flexible tube or channel. This squeezing action creates a wave-like motion that mimics the peristaltic motion observed in our digestive system, effectively moving the fluid forward.
3. Gear Micropumps: Gear micropumps are also widely used and utilize rotating gears or rotors with precise clearances to capture and transport the fluid. As the gears rotate, pressure differentials are created, enabling the fluid to be efficiently transported through the pump.
4. Piston Micropumps: Piston micropumps rely on a reciprocating piston to generate fluid movement. The piston displaces the fluid by retracting and extending within a chamber, resulting in pressure changes that propel the fluid flow.

For instance, a reciprocating diaphragm is employed as a physical stimulus to pump fluid [6]. These micropumps, often referred to as displacing micropumps, include two main elements for liquid propulsion and cessation, leading to a pulsating flow owing to their periodic nature [1, 6]. The mechanical/displacement pump consists of an input and outlet, a flexible diaphragm, an actuator, and a pumping chamber [6, 35]. Figure 4.1 illustrates the schematic operation of a mechanical micropump. An actuator is used to generate the required pressure differential for fluid pumping in a mechanical micropump, leading to oscillation or reciprocation. When the diaphragm oscillates, the chamber volume increases, causing the pressure inside the pumping chamber to decrease. Consequently, in the supply mode, the fluid flows from the high-pressure reserve to the pumping chamber due to the pressure decrease. Conversely, as the diaphragm flexes back and the chamber volume decreases, the chamber pressure increases. This signifies that the pumping mode is active, and the fluid exits the chamber through the outlet [6]. The pressure is a result of the actuator's stroke volume (ΔV) within the chamber. The actuator must work in conjunction with the dead volume (V_0) of the chamber. In micropumps with mechanical diaphragms, the compression ratio is an essential parameter. The compression ratio is defined by the following equation:

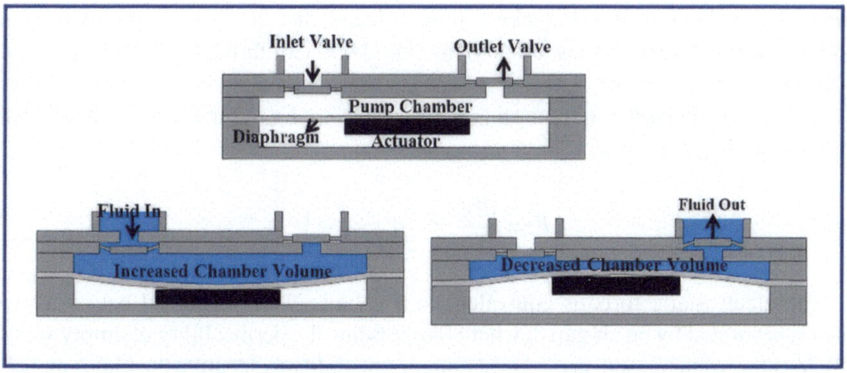

Fig. 4.1 Diagram demonstrating mechanical micropumps operating principle. Reprinted with permission from Ref. [6]. Copyright 2019, Elsevier

$$\varepsilon = \frac{\Delta V}{V_0}$$

Typically, the efficacy of mechanical micropumps is constrained by their mechanical components [7]. In mechanical micropumps, flow rectification can be achieved using microvalves such as ball valves, check valves, and other types, or through fixed geometry valves like diffusers or nozzles (valveless micropump) [6, 15, 29].

4.4 Comparison of Mechanical Micropumps Working Principles, Advantages, and Limitations

4.4.1 Electrostatic Micropump

The electrostatic actuation mechanism-based micropump is the second major type of mechanical micropump [9, 10]. Electrostatic excitation relies on the Coulombic attraction of two oppositely charged substances [5]. An electric force F attracts or repels the membrane or diaphragm connected with the electrostatic micropump; this mechanical movement is perceived as the pumping action [9, 10]. Two parallel plates, similar to capacitors, are applied to AC voltage sequentially in such a situation. In most cases, one can move while the other remains stationary. The schematic of a micropump operated by electrostatic force is shown in Fig. 4.2. Since no static current runs between the two plates, the electrostatic force emerges as the voltage is applied between electrodes. This causes the moving plate to deform in an upward direction, causing the chamber to expand. As a result, the import valve opens, allowing fluid to enter from the import. When the applied voltage is turned off or reduced, the deformation of the moving plate (membrane) is restored by the membrane's elastic force. The volume of the chamber will be reduced, and the export valve will open when the import valve closes. Finally, the liquid will exit through the outlet [15, 29, 36, 37].

The following equation represents the electrostatic force produced by an applied electric field E [3, 9, 10].

$$F = \frac{dW}{dX} = \frac{1}{2}\left(\varepsilon A \frac{v^2}{X^2}\right)$$

The electrostatic force is equivalent to a voltage V implemented between two planes separated by an air gap d, where 0 represents the permeability of empty space [5]. V is the voltage used, and ε is vacuum permeability to separate the plates and air or vacuum; its average value is typically equal to ε = 8.85.

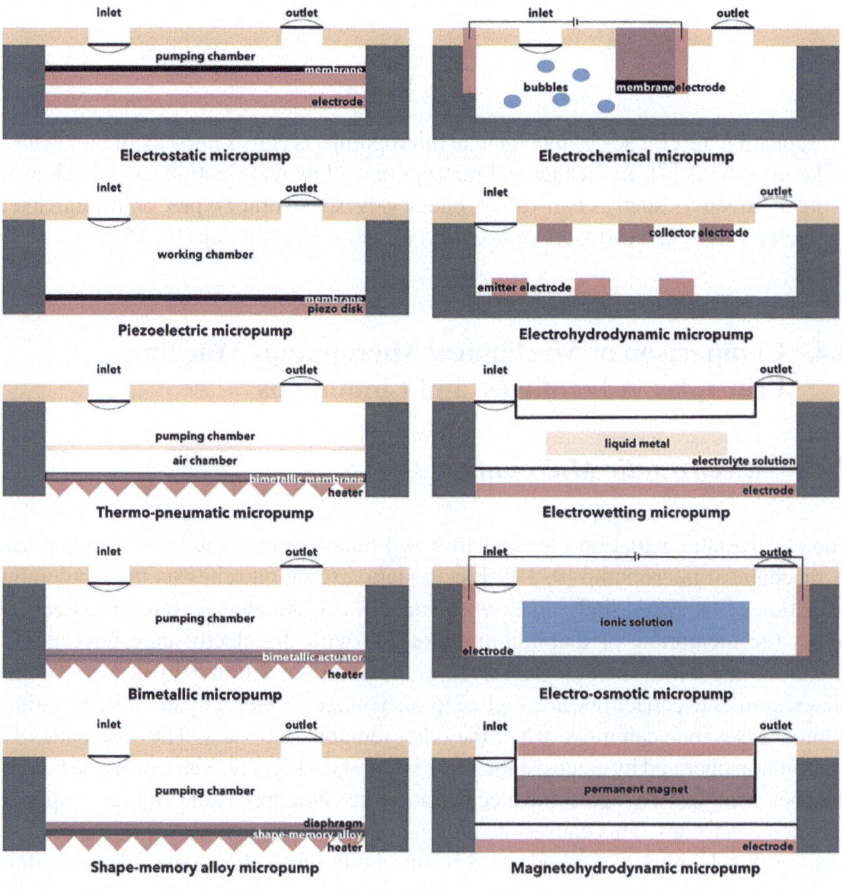

Fig. 4.2 Schematic representation of the workings of micropumps [38]

Electrostatic actuation is perfectly fit with MEMS microfabrication methods because of its fast response times and low energy loss. The main disadvantages of this method are the high actuation voltage (approximately 200 V) and the slight membrane displacement, which is typically limited to roughly 5 m [3, 15, 29, 37].

Many MEMS structures employ electrostatic deformation or diaphragm or beam pull-in. In biomedical applications, the capacitive pressure sensor and the capacitive accelerometer are commonly used for detecting blood pressure gradients in the heart's coronary artery. Actuating devices driven by electrostatic force include microelastic joints, Comb drivers, MEMS power switches, micropumps, and rotary electrostatic actuators. In a study by Kim et al., they proposed electrostatic peristaltic multistep micropumps with active microvalves. These micropumps achieved high airflow rates of 4.0, 3.0, and 2.7 cm, respectively, making them suitable for compressible gas pumping at high pressure and low power consumption. Moreover, adding functional valves (19 valves in the 18-step micropump) makes the manufacture and operating sequence more difficult. Kim's group created a miniaturized circulatory pillar method for gas chromatography uses was created using an electrostatic peristaltic micropump with 6 valves and 2.25-cm microcolumns. The circulatory column system was designed to prolong the useful length of the 2 columns by up to 8 m, resulting in a considerable increase in the goal cut method efficiency [39].

Furthermore, it is worth noting that the circulatory column system required a mere 5.5 kPa of head pressure, a value that is readily achievable with standard electrostatic peristaltic micropumps boasting a maximum capacity of 17.5 kPa. Electrostatically actuated micropumps and dielectric elastomer actuator (DEA) micropumps employ comparable operational mechanisms. A sinusoidal voltage is applied between a stationary layer and a flexible film, resulting in alternating liquid flows into and out of the pumping chamber. Ghazali et al. presented a micropump consisting of a pre-stretched dielectric elastomer actuator (DEA) placed between two compliant and flexible electrodes, with a pull-up spring positioned on the upper side [40]. The elastomer membrane initially has a flat shape, as shown in Fig. 4.3a. However, when a significant electrical potential difference is supplied between the electrodes, the dielectric elastomer actuator (DEA) undergoes contraction and expansion as a result of the Maxwell tension generated by the compressive electrostatic force. The device achieved the highest flow rate of 42 L/s at a 3 Hz actuation frequency due to an optimal membrane diameter and thickness. Electroactive polymer matter is gaining popularity for micropump uses due to its capacity to deform significantly in the presence of a weak applied voltage and to function in both aqueous and air conditions. Electroactive polymer composite membranes, unlike electrically actuated membranes, deflect due to ionic movement produced by introducing an external electric field. Ionic polymer-metal or Ionic conductive polymer film (ICPF) composite membranes are the most common polymer composite membranes (IPMC). A typical IPMC membrane, as illustrated in Fig. 4.3b, is composed of a coating of Silica/Nafion chemically, and perfluorosulfonic acid polymer has been plated with high electron density as electrodes, such as Pt or Au. IPMCs were widely used in micromanipulators, micro-robots, and medical equipment since they are simple to bend bidirectionally using intermittent electrical power.

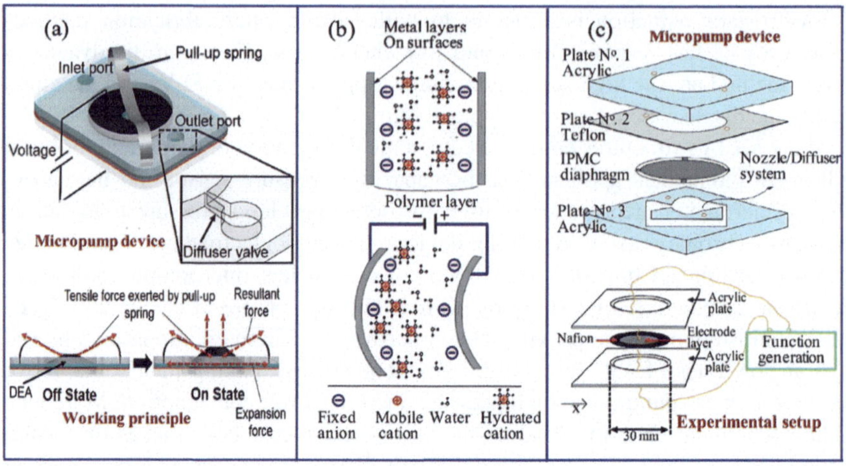

Fig. 4.3 **a** The dielectric elastomer actuator-equipped micropump's mechanical construction. **b** The IPMC actuator's operation. **c** The structural layout of the micropump with IPMC diaphragm (top) and the setting for the experiment (bottom). Reprinted with permission from Ref. [5]. Copyright 2018, Elsevier

Santos et al. assessed the Nafion–Pt polymer actuators' stroke volume at various actuation current levels (Fig. 4.3c). IPMC has an internal low-pass filter, therefore low-frequency operation (< 0.5 Hz) is necessary for the diaphragms. A high current flow of about 8.02 μL/s was achieved with a stroke volume of up to 80 μL at an operating frequency of 0.1 Hz. On the other hand, the manufacturing procedure was difficult since it required intricate Pt electrodes on the Nafion film patterning. In contrast to conventional designs, which included a solitary IPMC membrane along all edges, Nam et al. suggested several IPMC membranes arranged in a stacked configuration and secured at a single edge. Removing the edge limitations led to a substantial increase in membrane deformation, resulting in a much larger stroke volume [41]. In addition, the IPMC was separated from the pumping fluid in the proposed design, significantly enhancing the device's operational robustness. According to the experiments, the multi-substrate design extended the steady operating period from eight minutes to about thirty-five minutes [5, 42].

4.4.2 Piezoelectric Micropump

The piezoelectric diaphragm mechanism is the most often used kind of mechanical micropump [35]. A diaphragm-mounted piezoelectric disc, valves, and a pumping chamber make up a piezoelectric micropump. The conversion of piezoelectric materials activates the piezoelectric micro pump. An applied electric field on a piezoelectric crystal produces tension, as seen in the schematic depiction of piezoelectric actuation in Fig. 4.2 [28]. Piezoelectric actuators are commonly used in drug delivery systems employing membrane micropumps. The microscopic basis of piezoelectricity relies on the asymmetrical ionic charge transport. When an applied electric field induces dimensional stresses and changes in a piezoelectric material (inverse piezoelectric effect), it results in electric polarization (direct piezoelectric effect). Natural piezoelectric crystals such as quartz and zinc blends exhibit these properties. Modified lead zirconate titanate (PZT) ceramics are the most popular materials for piezo actuators [3, 43].

Piezoelectric pumps have widely used since the 1990s [8, 44]. Initially, piezoelectric disks were employed in inkjet print heads in the 1970s and later in MEMS micropumps by Smits and van Lintel [3, 45]. Piezoelectric actuators offer several advantages such as lower power consumption, simple and compact structures [46], immunity to magnetic fields, fast response, and high displacement resolution, and are widely utilized in the fields of ultra-precision engineering and micro actuators.

However, the fabrication process is challenging due to the difficulty of working with piezoelectric materials [15]. While piezoelectric actuation requires less energy, the operating voltage is often high in most applications, exceeding 100 V. This high operating voltage limits the use of piezoelectrically actuated micropumps. Moreover, mass manufacturing using piezoelectric actuation techniques is challenging, and the production costs are expensive, which raises concerns about reliability [36]. Before piezoelectric micropumps can be widely used in medical applications, various obstacles need to be overcome. Disposable flow channels are employed to reduce contamination and infection risks. Additionally, the detachability of micropump actuators and other components allows for quick repair of worn or broken flow channels. To achieve a more cost-effective solution, one option would be to replace only the chamber while retaining the piezoelectric actuator. However, this approach is not common in most piezoelectric micropump designs. Typically, the entire pump needs to be discarded after each use. Therefore, it is essential to focus on designing a piezoelectric actuator that can be isolated from the pump chamber. This approach holds the potential for reducing both expenses and waste. Researchers have developed detachable thermo-pneumatic micropumps and peristaltic designs for detachable piezoelectric micropumps [46]. However, peristaltic pump systems with control circuits and cascade chambers are more expensive compared to single-chamber pumps.

The relationship between electric field (E), electric displacement (D), strain (S), and mechanical stress (T) according to piezoelectric coupling is provided by the following equations [28]:

$$[\mathrm{S}] = \left[S^E\right][\mathrm{T}] + [\mathrm{d}][\mathrm{E}] \tag{4.1}$$

$$[\mathrm{D}] = [d]^T[\mathrm{T}] + [\varepsilon][\mathrm{E}] \tag{4.2}$$

where d, D, T, SE, E, S, and ε represent the piezoelectric strain coefficient, charge displacement, surface stress, compliance with zero fields, electric field, strain, and dielectric constant of a piezoelectric material, respectively.

The actuation method of a piezoelectric material is determined by its crystal structure, where the “d” coefficient serves as the mechanism for converting one form of energy into another. The “d” coefficient, also known as the piezoelectric coefficient or piezoelectric charge constant, quantifies the material’s ability to generate an electric charge or voltage in response to applied mechanical stress or strain. This property allows piezoelectric materials to convert mechanical energy into electrical energy and vice versa. The actuation of the material is defined by three different modes: shear mode (d15), longitudinal mode (d33), and transversal mode (d31), depending on the directions of the electric field (E) and polarization (P). Figure 4.4 illustrates the piezoelectric actuation method with the six-coordinate system orientations (y, z, x, θy, θz, and θx) and the polarization of a unique substrate piezoelectric crystal under P. For d31 and (d33) modes, where E is applied parallel to P, transversal (δl) and longitudinal (δh) deformations co-occur (Fig. 4.4a, b). In contrast, for (d15) mode, where E is vertical to P, shear transformation (s) takes place (Fig. 4.4c). Among these modes, strains are greatest in (d31) and (d33) compared to d15. Piezoelectric actuators are widely used in high-precision actuation applications, such as micro-robotics, micro/nano-positioning systems, and micropumps. This is due to their ability to generate fast and precise small-scale deformations. However, the incorporation of piezoelectric materials like lead magnesium niobate-PZT ceramics and lead zirconate titanate (PZT) into MEMS fabrication is challenging due to the instability of deposited materials and the requirement for high-temperature thermal processes. Other factors that limit their use include the need for relatively high driving voltage and significant nonlinearity/hysteresis [47–49].

4.4.3 Thermo-Pneumatic Micropump

An important category within mechanical-type micropumps is the thermo-pneumatic actuation mechanism. This mechanism can utilize either a peristaltic movement or a vibrating diaphragm [35]. It’s crucial to note that the motive power is provided by the heated fluid, not the stimulated fluid [36].

In a thermo-pneumatic micropump, there is typically another chamber located above the pumping chamber, which includes a radiator (refer to Fig. 4.2). The working concept of the thermo-pneumatic actuator is based on the phase change or volume change of the liquid present in this upper chamber with a radiator [3].

These actuators are created by enclosing gases with suitable thermal expansion characteristics in a chamber that has a membrane on one side. The chamber also contains a heating element. When the heating element is activated, the volume of the gaseous content increases, leading to a rise in pressure within the chamber. As a result, the membrane undergoes deformation [28]. When the heating element is turned off and the air begins to cool, the volume of the air decreases. The ability of the membrane to deform and return to its original shape when the air is heated and cooled facilitates the pumping of medication into and out of the pump [51–54].

The change in pressure in a liquid can be represented by the equation:

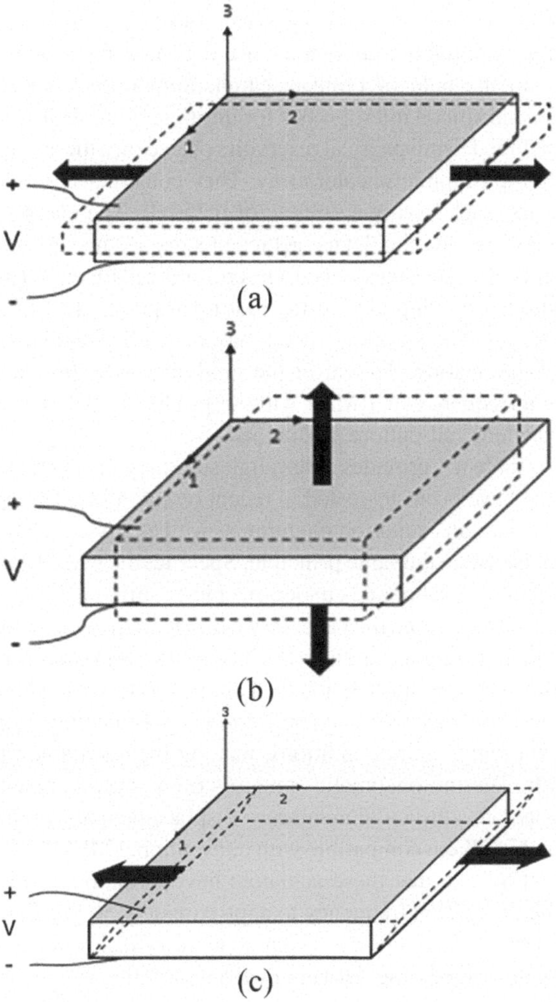

Fig. 4.4 Modes of piezoelectric performance: **a** the d31 mode, **b** the d33 mode, and **c** the d15 mode [50]. Copyright 2019, journals.sagepub

$$\Delta P = E\left(\beta \Delta T - \frac{\Delta V}{V}\right)$$

where ΔP is the pressure variation, ΔT is the temperature rise in the chamber, ΔV/V is the percentage change in volume, β is the thermal distension coefficient of the liquid, and E is the modulus of elasticity of the diaphragm material [5]. For simplicity, let's assume there is no volume change, and we'll use $\beta = 2.3 \times 10^4$ C^{-1} and $E = 3.3 \times 10^5$ psi for water in Eq. (8). Therefore, under these conditions, the temperature-dependent pressure change for water can be expressed as 76 psi/°C. Thermal actuation techniques, including thermo-pneumatic micropumps, are characterized by high pressure, significant power consumption, and a long reaction time.

While these micropumps generate substantial pressure and membrane displacement, it is essential to maintain the driving force above a specific threshold [15, 55, 56].

In the field of brain research, Spieth et al. developed a micropump utilizing special silicon microprobes for intra-cerebral medication administration. The device included hemispherical reservoirs placed on a thermally expanding polymer substrate on top of a microheater array. They constructed a model device with sixteen reservoirs, each having a volume of 0.25 μL. The energy consumption per dosage was 3.357 W, and the device achieved a dosage accuracy of 0.25 ± 0.02 μL [57]. Bonk and colleagues introduced a microfluidic platform for studying cellular activity. They utilized on-chip electro-thermal micropumps for drug administration and medium mixing. The pumping speed may be readily regulated by manipulating the voltage of the heating element or the field electrode. In addition, these pumps functioned at a frequency of 1 MHz without producing bubbles and were suitable for use with standard cell-culture media [58].

Table 4.1 provides a thorough summary of several thermal actuation micropumps that have been suggested in recent research [5]. The operation of these micropumps involves an excitation medium, as well as a phase change and pneumatic expression, all based on the same principle. Specifically, when heated, materials like paraffin and perfluoro compounds undergo a phase shift, enabling effective phase-change actuation. To generate the necessary heat, micro heaters made of Gold/Chromium (Au/Cr) and Chromium/Platinum (Cr/Pt) are deposited on a glass substrate using Electron Beam Evaporation. In medical applications where skin-contact actuation is required, materials such as methyl perfluoropropyl ether undergo a phase change under normal skin temperature conditions, making them suitable for powering micropumps [28, 59]. Thermo-pneumatic actuators offer several advantages, including the ability to generate substantial membrane displacement and high pressure at low input voltage, making them compatible with conventional MEMS processing. However, it's important to note that these actuators have a relatively slow reaction time, which limits their actuation frequency to approximately 50 Hz. Furthermore, their high electrical energy consumption is a significant drawback when considering their use in portable or implanted drug delivery devices [3, 59].

4.4.4 Bimetallic Micropumps

Bimetallic micropumps involve the use of objects made from two distinct metals, known as bimetals, which exhibit different coefficients of thermal expansion [7]. Thermal stresses are produced when these materials are connected and exposed to changes in temperature. Because the chosen materials have limited thermal expansion coefficients, bimetallic pumps have low diaphragm deflection despite the large forces acting upon them, leading to simple operation. Unlike other types of micropumps, bimetallic pumps operate at lower voltages but are not suitable for high frequencies. Figure 4.2 provides a schematic representation of a bimetallic micropump,

Table 4.1 Micropump devices for blood manipulation: features and performance

Device recommender	Device features	Device performance
Jebrail et al. [60]	The device uses a droplet-based microfluidic (DMF) method to extract and purify RNA from lysates of whole blood. It is made up of an extraction module and a DMF-based purification module joined by a bridge droplet. Analyte-binding magnetic beads in the droplet are released and rendered immobile by a moveable magnet positioned underneath the bridge electrode	The device achieves RNA extraction from functionalized particles using a peristaltic micropump connected to a wide-bore tube. Compared to standard procedures, this technique is more than two times faster and requires 12 times fewer chemicals, while maintaining the same RNA yield and quality levels
Feeny et al. [61]	This device incorporates an improved pumping mechanism designed for time-controlled pumping of dried samples on filters. It utilizes PDMS gas-permeability to facilitate fluid transport in the microfluidic channels	The device has been successfully used to extract dopamine in the range of 250 micromolar to 1 mM from dried filter samples within approximately 5 min
Zehnle et al. and Schwemmer et al. [62, 63]	The proposed technique involves a radial inward pumping mechanism onto a rotating microfluidic disk using an air-filled chamber	The micro pumping mechanism demonstrates a pump cycle efficiency of over 75% for various liquids, including whole human blood, highly viscous lysis buffer, ethanol, and water. This approach efficiently transfers a significant volume of liquid without requiring additional actuators or complex manufacturing processes, making it advantageous for more advanced tests that require higher fluidic throughput
Song et al. [64]	The microfluidic cell sorting system incorporates electromagnetic pumps, flow-focused channels, detecting channels, and a differential resistance pulse sensor (RPS). The size difference between target and non-target cells allows for identification. The RPS signal is used to direct a target cell into a collecting chamber and turn on the electromagnetic pump once the cell has been recognized	A549 cells and T-lymphocytes are effectively identified and sorted by the device at a rate of up to 100 target cells per minute from peripheral fingertip blood samples. The activation of the electromagnetic pump has little impact on the RPS cell identification and sorting procedures, making it appropriate for the sorting of infrequent target cells such as circulating tumor cells

(continued)

Table 4.1 (continued)

Device recommender	Device features	Device performance
Kant et al. [65]	This device utilizes a peristaltic micropump with fluid chambers specifically designed to minimize reverse flow and optimize discharge efficiency per unit volume. It is used for the transportation of blood samples	The peristaltic micropump effectively transports blood samples, ensuring efficient and controlled delivery
Dehghan Manshadi et al. [66]	To better understand and manipulate non-Newtonian liquids like blood, researchers used numerical simulations to investigate and improve the efficiency and voltage limitations of parallel electrostatic micropumps	Through analysis and optimization of parallel electrostatic micropumps' voltage limitations and performance, the simulations attempted to enhance the handling of non-Newtonian liquids such as blood
Kumar et al. [67]	The device is a planar valveless micropump controlled by a solenoid. It is used for the transfer of diluted blood samples and biological cell lines, offering multiple/single input output configurations	Based on experimental results, the device is capable of achieving a maximum flow rate of 94 µl/min for HL60 cells and diluted blood samples when operated at a frequency of 90 Hz. At a higher frequency of 135 Hz, the device can reach a flow rate of 106 µl/min for the same samples. The planar valveless micropump, controlled by a solenoid, enables efficient transfer of diluted blood samples and biological cell lines in various configurations

featuring a diaphragm composed of two different metals that exhibit varying distortions when heated. The thermal variation induces diaphragm deviation due to the different thermal expansion coefficients of the chosen materials [15, 33].

The volumetric thermal expansion coefficient (αv), as defined in Eq. (4) [35], is relevant for fluid applications:

$$\alpha_v = \frac{1}{V}\left(\frac{\partial V}{\partial T}\right)_P$$

Here, "T" and "V" represent temperature and volume, respectively, while "P" denotes constant pressure during thermal expansion. In a bimetallic micropump, the metal film changes its path based on its thermal expansion coefficient in response to the temperature difference. This diversion of the metallic film leads to a pressure shift in the air chamber, causing deflection of the elastic membrane. The degree of deflection, which results from the pressure change in the air chamber and the deviation of the metallic film, can cause either an upward or downward deflection. In response to changes in pressure within the pumping chamber or container chamber, the inlet and outlet valves adjust their opening and closing positions [35]. Bimetallic micropumps utilize materials with low thermal expansion coefficients, resulting in minimal deflection in the bimetallic actuation system [7, 68].

4.4.5 Alloy for Shape Memory

Shape memory alloy (SMA) micropumps, shown in Fig. 4.2, are devices that use shape memory alloys to stimulate fluid flow on a micro-scale. (SMA) is a unique material that can return to its original shape when exposed to specific stimuli, such as heat or magnetic fields. This property makes SMA an attractive choice for actuation mechanisms. The shape memory effect in SMA involves a phase change between two solid phases: austenite at high temperatures and martensite at lower temperatures. The martensite phase is more flexible than austenite, allowing for significant shape changes in SMA materials [28]. When the SMA material is heated above its austenite start temperature, it undergoes a phase transition and reverts to its pre-deformed shape. However, if mechanically constrained, it exerts force while maintaining its deformed shape. This mechanical distortion is harnessed for actuation purposes, but it requires a substantial amount of electricity and has a slow response time [15]. The most commonly used SMAs are Nickel-Titanium (NiTi) and Copper–Aluminium-Nickel alloys [28]. SMA materials, due to their low weight and high biocompatibility, are well-suited for Micro-Electro-Mechanical Systems (MEMS) applications. They find applications in fields such as medication delivery and chemical resistance. However, challenges arise from uncontrolled transformations during cooling or heating processes and high energy consumption, limiting the practical use of SMA-based devices [9, 10]. Temperature sensitivity, high power consumption, and low actuation frequency are the main disadvantages that affect precise control of SMA deformation [3, 69].

In SMA micropumps, a nickel/titanium alloy diaphragm (TiNi) is commonly used as the activator. TiNi offers high actuation forces and recoverable strain, enabling the micropumps to achieve high operating pressures and pumping rates. One such micropump was reported by Benard et al. (1997), which used a silicon spacer to separate two TiNi membranes. The design included fixed check valves and a cantilever for flow correction. The reciprocating motion was achieved by alternately heating the two TiNi membranes using joule heating. This micropump achieved a maximum flow rate of 49 l/min at a function frequency of 0.9% and a back pressure of 4.23 kPa. It operated at a working current of 0.9 A, 6 V voltage, and consumed 0.50 W of power. Another design by Xu et al. featured a polyimide spring-biased SMA micropump with lower flow rates. Its overall dimensions were approximately 6 mm by 5 mm by 1.5 mm, and it utilized free-standing SMA thin films to achieve a volumetric flow rate of 340 l/min at 100 kPa back pressure. This micro SMA pump required a unique bias mechanism and structure to isolate the working liquid from the driving circuits. However, an a-Si/NiTi composite driving membrane reported in another reference eliminated the need for a specific bias structure. The silicon layer provided the necessary biasing force, simplifying the micropump's construction and achieving high flow rates, good drive performance, and extended service life [70–72].

For use in biomedicine, Fukuda and Shuxiang created a SMA-actuated micropump. The components of this micropump were an elastic tube pump chamber, a casing, diffusers acting as one-way valves, and an SMA winding actuator. The actuator made use of an SMA coil, more precisely a 0.2 mm-wide TiNi cable. This micropump has a frequency-adjustable flow rate range of 500–700 l/min. It revealed prospective uses in biotechnology and medical operations, including intra-cavity involvement in surgeries and diagnostics, and displayed microflow capabilities [15, 73].

4.4.6 Electromagnetic Micropumps

Electromagnetic micropumps are a type of drug delivery system used in Microelectromechanical systems (MEMS). These systems are typically powered by embedded batteries, which limits their miniaturization due to the size constraints of the battery and electronics. To address this limitation, researchers have explored the development of magnetic force-controlled medication delivery systems [8]. An electromagnetic (EM) or magnetostrictive mechanism is used in micropumps with magnetic actuation [5]. These micromagnetic devices are made up of permanent magnets or powered coils with currents that trigger soft magnetic cores [15]. The diaphragm reciprocates due to the Lorentz force when a low-frequency square-wave electric current is supplied to a network of microwires and two permanent magnets placed on opposite sides of the chamber produce a transverse uniform magnetic field [74, 75]. You may express the Lorentz force (Fl) as follows:

$$F_l = n(I \times B)$$

where n is the number of parallel microwires, B is the external magnetic flux density, I is the current, and L is their length [5, 76, 77].

Electromagnetic activation, using copper and NiFe, is not commonly used in micropumps. When current is applied, a silicon sheet is deformed by magnetic force, as shown in Fig. 4.5a. Despite its low operating voltage, electromagnetic actuation consumes a significant amount of electricity. Producing electromagnetic components using standard MEMS techniques is also challenging. In addition to electrowetting, shape memory alloys, and electrohydrodynamic actuation, other actuation techniques have been utilized. However, there are fewer practical applications of the aforementioned actuation techniques [36]. The diaphragm of an electromagnetic (EM) micropump moves due to the interaction between permanent magnets and a varying magnetic field produced by a micro-coil carrying an electric current. The magnetic actuation force may exhibit either repulsion or attraction, enabling the membrane to deflect in both directions by manipulating the phases of the input current. Chee et al. created an electromagnetically operated micropump (as seen in the Fig. 4.5b) by using a flexible PDMS membrane, a chamber with a gourd-like form, and a solenoid plunger. The membrane is asymmetrically deflected when the plunger is activated, which creates a pressure difference within the chamber and permits pumping activity. The micropump reached peak flow rates of 1.48 ml/min in the reverse direction and 1.52 ml/min in the forward direction when it was operated at resonance frequencies of 20 and 15 Hz. PDMS-based micropumps offer numerous advantages, but their manufacturing process is generally complex. It often involves bonding and casting multiple PDMS substrates, and any errors in these steps can jeopardize the microstructures created during the initial replication phase and impair the pump's functioning. Furthermore, the diaphragm of the majority of micropumps made from PDMS is activated by exerting an external force, such as a magnetic field or altering the pressure of the working fluid, to displace magnetic materials that are in contact with it [5]. Electromagnetic actuators, due to their benefits such as significant deflection, rapid reaction time, high field energy density, and low input voltage requirements, have been employed in various MEMS applications. However, they also have drawbacks, including parasitic loss at high frequencies, high power dissipation for driving coils, and a decrease in produced electromagnetic forces as the device size shrinks, which must be considered in device design [29, 47, 78].

Pir Moradi did innovative research on a micropump design that used magnetic PDMS sheets with laser-drilled perforations to seal drug-filled micro-reservoirs. The activation of these reservoirs was achieved by the application of an external magnetic field (Fig. 4.5c). By incorporating iron oxide nanoparticles covered with a surfactant into the polymer, Pir Moradi achieved magnetism in the PDMS membranes. The PDMS material was shaped into drug reservoirs using photolithography techniques, and a sacrificial coating of polyacrylic acid was utilized to attach the reservoirs to the magnetic PDMS membrane. The magnetic composite membrane was deposited through spin-coating. To enable the filling of drug solutions into the reservoirs, Pirmoradi modified the hydrophobic PDMS surfaces with bovine serum albumin to improve surface wettability. As a result of applying an external magnetic field, the membrane bent concavely, leading to increased reservoir pressure and controlled medication release. Initially, Methylene Blue was chosen as the drug due to its high solubility and UV-V absorptivity (Fig. 4.5d). Pirmoradi employed a rotating permanent magnet to provide the magnetic field for actuating the cell walls. The amount and duration of medication released were effectively controlled by adjusting the number of actuation events and the amplitude of the external magnetic field (Fig. 4.5e). Furthermore, Pirmoradi successfully utilized the micropump to deliver docetaxel, a chemotherapeutic medication with antiangiogenic effects in treating diabetic retinopathy. For almost a month, the micropump maintained a steady and repeatable release rate, with hardly little medication leaking out while the device wasn't in use. Importantly, thorough in vitro investigations conducted by Pir Moradi's group verified the released drug's biological activity [8, 79].

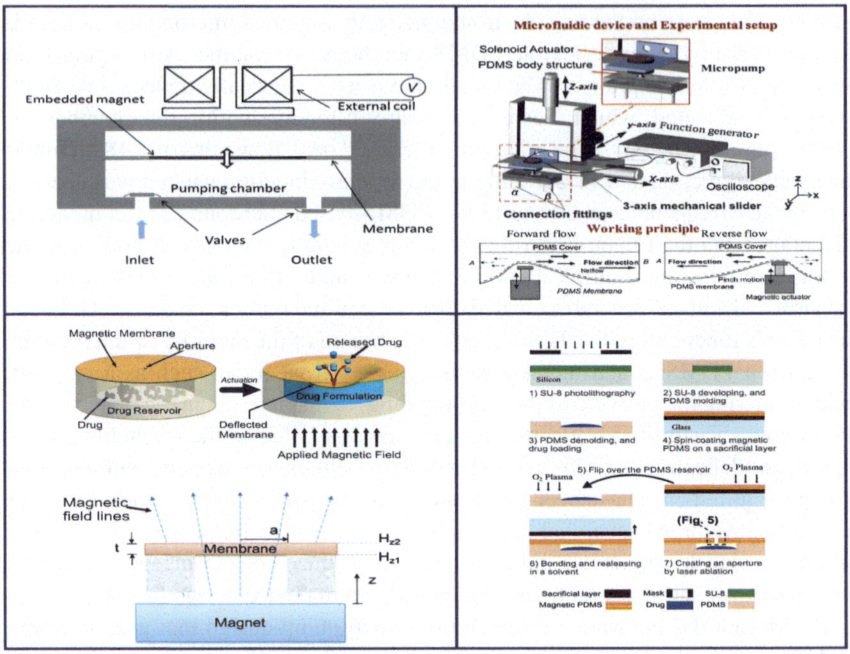

Fig. 4.5 **a** Electromagnetic actuation schematic view. **b** An electromagnetic micropump schematic (upper) shows the micropump's operating principle in forward and reverse modes (lower). Reprinted with permission from Ref. [5]. Copyright 2018, Elsevier. **c** Operating principle, (1) prior to actuation, (2) drug release following the application of a magnetic field. **d** A permanent magnet is used to activate the incitement mechanism of a hung magnetic membrane in an outer magnetic field. **e** A magnetically actuated drug delivery device's major manufacturing stage. Reprinted with permission from Ref. [8]

4.4.7 *Ion Conductive Polymer Film (ICPF) Micropumps*

Ion conductive polymer film (ICPF) micropumps are microfluidic devices that utilize ICPF actuators to generate fluid flow at a small scale. These micropumps are designed for handling small volumes of liquid and have applications in biotechnology, medical devices, and lab-on-a-chip systems (Fig. 4.6a). ICPF micropumps consist of a pump chamber and one or more ICPF actuators, which act as the diaphragm or membrane of the pump. The ICPF actuators are responsible for actuating the diaphragm, creating the required pressure to propel the fluid. The pump chamber is typically equipped with one-way check valves that ensure fluid flow in the desired direction and prevent backflow [15]. To activate ICPF micropumps, appropriate voltages are given to the ICPF actuators. The application of an electric field induces the motion of ions inside the ion-conductive polymer film, resulting in the expansion or contraction of the ICPF actuator and the consequent deformation of the diaphragm. This deformation generates pressure within the pump chamber, resulting in fluid flow through the check valves (Fig. 4.7b). ICPF micropumps offer several advantages, including low power consumption. The actuation of ICPF actuators requires relatively low voltages, making them energy-efficient compared to other micropump types. Moreover, the biocompatibility of the ion-conductive polymer film makes these micropumps suitable for biomedical applications, such as drug delivery systems and microfluidic diagnostic devices. These micropumps provide precise control over fluid flow rates and can achieve high pumping efficiencies. They can be integrated into compact and portable systems, making them ideal for point-of-care diagnostics and wearable medical devices. However, the manufacturing process of ICPF micropumps can be complex due to the fabrication of ion-conductive polymer films and the integration of the pump chamber and check valves [15]. It is worth noting that ICPF actuators are commonly used in polymer-based MEMS devices. Two thin metal electrodes, such as gold and platinum, are placed over a layer of perfluorosulfonic acid polymer to create an anionic conductive film (ICPF) for use in membrane micropump actuators [3]. Physically, the ICPF actuator resembles a diaphragm between two thin films of the polymer. The electrical conductivity of these two layers is quite high. One end of the diaphragm is stabilized, and the ICPF diaphragm can be bent in either an upward or downward direction by applying suitable voltages to the electrodes. Because of its low actuation voltage, biocompatibility, and high bending displacement, the ICPF actuator is frequently referred to as an artificial muscle [10, 80].

The micropump's diaphragm is formed by integrating the two one-way check valves and the pump chamber with ICPF actuators. For high flow rates, ICPF actuators are mounted in series. Studies have described the use of ICPF in robotics, micromanipulators, and medical devices. Guo et al. have described the development of ICPF micropumps for biomedical applications based on polymer actuators. The micropump design has potential applications in biotechnology and medical fields due to its sufficient flow rate, biocompatibility, and low power consumption. Guo and colleagues have also described other applications of ICPF actuators, such as artificial fish microrobots with potential medical applications, including conducting

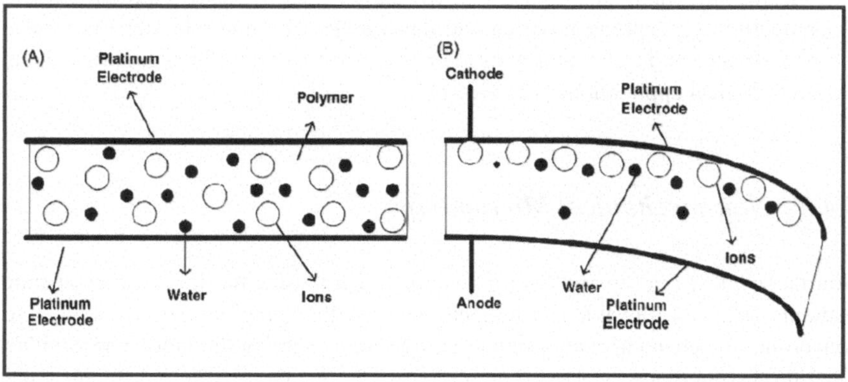

Fig. 4.6 **a** Shows a schematic representation of the structure of an ICPF (Ionic Conducting Polymer Film), (**b**) and the bending principle of an ICPF [15]

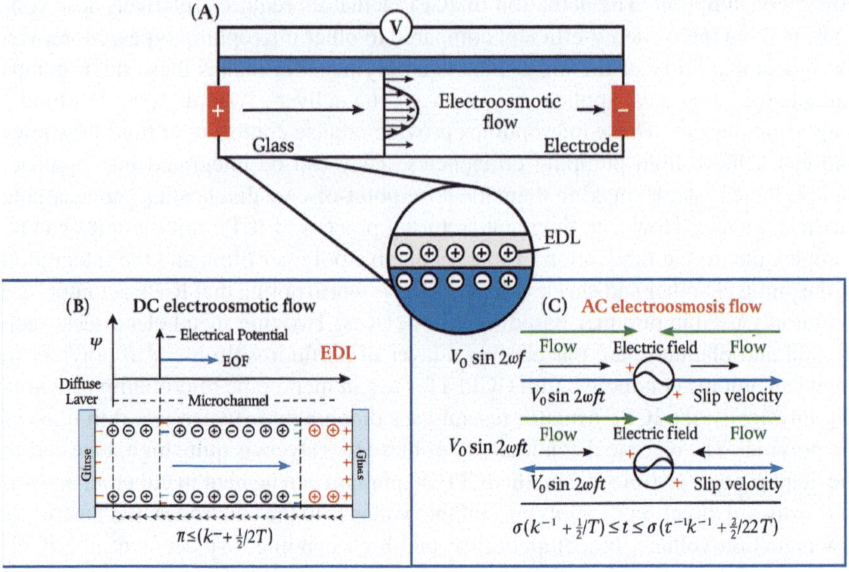

Fig. 4.7 **a** Electroosmotic micropump schematic. An EDL is formed spontaneously by the charges in the ionic solution and the charges in the glass tube. The electric field used along the tube will cause the mobile ions in the EDL to moving, pulling along the other liquid particles, resulting in a net flow. Reprinted with permission from Ref. [3]. **b** Flow of electroosmotic DC. **c** Flow of electroosmotic AC. Reprinted with permission from Ref. [5]. Copyright 2018, Elsevier

delicate surgical operations to minimize needless incisions. The advantages of the ICPF actuator include low driving voltage, fast response, and biocompatibility. It can also operate in aquatic settings. The manufacturing of the ICPF actuator is complex, which is a key constraint. Overall, ICPF micropumps offer a promising solution for efficient fluid manipulation at the microscale, with low power requirements and biocompatibility. Ongoing research and development in this field aim to enhance the performance and functionality of ICPF micropumps for various biomedical and biotechnological applications [15, 33, 81].

4.4.8 Non-mechanical Micropumps

Non-mechanical micropumps are a category of microfluidic devices that operate without the use of movable components. Instead, they employ various methods to convert non-mechanical energy into kinetic energy, enabling fluid movement within microfluidic components. These micropumps can be classified into different types based on their underlying principles and mechanisms. Non-mechanical micropumps offer several advantages over their mechanical counterparts. They often have simpler designs, smaller form factors, and lower power consumption. Moreover, they may be seamlessly incorporated into microfluidic systems for many purposes, including point-of-care diagnostics, lab-on-a-chip devices, drug delivery systems, and precise manipulation of small-scale fluids for chemical analysis and biological experiments. However, non-mechanical micropumps also have limitations. They typically exhibit lower flow rates compared to mechanical pumps, and their performance can be influenced by factors such as fluid properties, channel geometry, and surface interactions. Moreover, precise control of operating conditions, such as voltage or temperature, is often required to achieve desired flow rates and pump functionality [82, 83].

4.4.9 Electrochemical Micropumps

Electrochemical micropumps operate by using reversible electrochemical reactions (Fig. 4.2) [35]. These micropumps rely on gas bubble generation through the electrolysis of water, which differs from conventional pumping mechanisms where electrolysis is typically an undesired side effect. In electrochemical micropumps, electrolysis serves as the primary driving force. Direct current (DC) is applied through two electrodes, usually made of platinum, immersed in a diluted electrolyte solution [3]. This process leads to the production of dihydrogen at the positive electrode and dioxygen gas at the negative electrode. The expansion volume resulting from the conversion of fluid water to gaseous dioxygen and dihydrogen is significantly higher at chamber temperature, allowing for substantial mechanical displacement and pumping action [3]. To create electrochemical micropumps, integrated electrodes are fabricated on silicon wafers using photolithography and lift-off processes. Titanium/platinum electrodes have been commonly used, deposited using techniques such as electron beam evaporation or magnetron sputtering [8]. The micropump structure consists of liquid channels, power supply electrodes, inlet and outlet tanks, and an electrolysis chamber for bubble production. Combining electrochemical micropumps with other microfluidic systems is straightforward due to their simple construction and design. However, it is important to note that bubble bursting in electrochemical micropumps can lead to unreliable and unstable drug delivery [15, 84, 85]. A specific electrochemical micropump design featured metallic electrodes on a polydimethylsiloxane (PDMS) structure with a glass substrate and electrolyte channels. The PDMS structure was created using soft lithography, while deep reactive ion etching (RIE) was utilized for diffusers and channels. During initial testing, two major issues were identified. First, the electrodes deteriorated gradually after several hundred cycles of alternating polarization. Second, the measured flow rate was lower than expected, which could be attributed to PDMS material relaxation and gas expansion in the chamber rather than efficient pumping. To address these issues, biodegradable materials were suggested for the electrodes, and a more rigid micro-pump chamber was proposed [86].

A new prototype was developed to test these improvements, using titanium/aluminum (Ti/Al) and platinum electrodes. Intermittent polarization resulted in an increased pump frequency due to a shorter gas shut-off period. However, because of PDMS deformation, the present rate continued to be lower than expected. Aluminum electrodes covered in a titanium coating degraded less quickly than platinum electrodes, which had higher breakdown rates. Intermittent polar short-circuit pulses generated ephemeral bubbles, which, upon ignition, provided sufficient energy to operate the pump. This regime resulted in significantly higher current measurements compared to the absence of ephemeral bubbles [8]. Further investigations focused on understanding the dynamics of gases in the chamber. Researchers discovered that using a larger voltage range resulted in a shorter pump duration. The cylindrical symmetry of the electrodes allowed them to explode independently of the dispersion of nanoscale bubbles in the chamber. Enhancing performance qualities involved selecting appropriate enclosure materials and utilizing aluminum electrodes. Microbubbles generated by explosion showed great potential for applications in various electrochemical actuators and pumps, including those with bubble force-activated membranes [87–89]. In another study, researchers designed and tested an electrolytic activator-based microelectromechanical systems (MEMS) device for drug delivery. The device consisted of two chambers: a pump chamber and a drug reservoir with gold interdigitated electrodes. The fabrication process involved electron-beam evaporation of titanium and gold layers on silicon, followed by assembly using polydimethylsiloxane (PDMS). PDMS was chosen for its flexibility, cost-effectiveness, biocompatibility, ease of fabrication, and transparency. The drug delivery system utilized doxorubicin and sterile deionized water as the electrolyte. The device demonstrated controlled drug delivery and successfully inhibited the proliferation of pancreatic cancer cells in vitro [8, 90].

In summary, electrochemical micropumps harness reversible electrochemical responses to generate gas bubbles for pumping. These micropumps offer potential applications in drug delivery and other microfluidic systems. However, challenges such as electrode degradation, lower-than-expected flow rates, and bubble bursting need to be addressed for improved performance and reliability.

4.4.10 Electrohydrodynamic Micropumps

Electrohydrodynamic micropumps rely on the interplay between electrostatic forces and ions present in a dielectric liquid to achieve fluid pumping (Fig. 4.2) [35]. These micropumps are specifically designed to operate with fluids that possess poor conductivity and dielectric properties. The conductivity of the fluid typically falls within the range of 10^{12}–10^{6} S/m [3]. The functionality of an electrohydrodynamic (EHD) micropump can be classified into three main types: conduction, induction, and injection. In the injection pump configuration, the electrodes within the pump generate electric charges, whereas the induction principle governs the operation of the inductive pump. The polarization pump utilizes the Kelvin polarization force to induce fluid movement [91].

The electric field generated by the EHD micropump corresponds to the energy density Fe and is denoted as the magnitude of the electric field E. It can be expressed as follows [91]:

$$F_e = qE + P\nabla E - \frac{1}{2}E^2\nabla\varepsilon + \frac{1}{2}\nabla\left[E^2\left(\frac{\partial\epsilon}{\partial\rho}\right)_T\rho\right]$$

In this equation, P represents the polarization vector, T denotes the temperature of the liquid, ρ stands for the liquid density, ε represents the liquid permittivity, and q represents the charge density.

These parameters and forces play a crucial role in the functioning of electrohydrodynamic micropumps, allowing for controlled fluid flow and manipulation within microscale channels. Ongoing research and development in this field aim to optimize the design and performance of EHD micropumps, opening up possibilities for various microfluidic applications [35, 91].

4.4.11 Electroosmotic (EO) Micropumps

Electroosmotic (EO) micropumps, along with electrophoresis (EP) and dielectrophoresis (DEP), are examples of electrokinetic phenomena commonly used in microfluidic systems. EO micropumps specifically employ electroosmosis to facilitate the pumping of electrolyte solutions. In EO micropumps, an electric field applied externally causes the anionic solution to move in proportion to the charged and stationary surfaces. This phenomenon occurs due to the charging of solid channel surfaces when in contact with an aqueous solvent, even if they are initially uncharged. The electrodes in EO micropumps are typically attached to fused silica or glass capillaries. When a silica surface with silanol groups is immersed in an aqueous solution, the solvation of the alcohol group's hydrogen atom negatively charges the solid surface, resulting in the formation of an electric double layer (EDL). By applying an electric field in these conditions, the cations within the EDL, such as those on a silica surface, migrate in the direction of the electric field, while anions move in the opposite direction. The resulting drag force at the surface generates a pure flow of the bulk fluid. Notably, electroosmotic flow exhibits a flat velocity profile (shown in Fig. 4.7a) [3, 92–94].

Both AC-operated and DC-operated fields can be utilized in the creation of EO micropumps (Fig. 4.7b, c). While most EO micropumps operate on a constant DC voltage, issues such as pH fluctuations and gas bubble formation due to electrolysis at the electrodes can arise in microfluidic channels. These problems can be mitigated by applying an alternating AC voltage. However, using AC voltage does not result in a pure electrolysis or a pure flow, making it more suitable for mixing than pumping. To generate a pure electroosmotic flow, nonuniform electric fields using asymmetrical electrodes or moving electric fields can be employed. It should be noted that AC-operated EO micropumps are not as widely used and have lower pumping strength compared to DC-operated pumps. The lack of moving parts in EO micropumps makes them ideal for handling delicate biologics. Scaling down the size of the device offers advantages such as lower voltage requirements, reduced sensitivity to Joule heating, and the ability to achieve smooth speed profiles. However, the performance and reliability of EO micropumps can be influenced by how fluids interact with surfaces, which is a concern in various fluid applications. Limitations of EO micropumps include the need for an electrically conductive solvent and high voltage. Despite these limitations, EO micropumps offer several benefits. They do not require moving components like check valves, and they can be manufactured using standard and cost-effective microelectromechanical systems (MEMS) methods. The direction of fluid flow in EO micropumps can be controlled by changing the direction of the external electric field. Due to their capabilities, EO micropumps find applications in chemical analysis, biochemical research, environmental monitoring, pharmacology, forensics, and clinical sciences [15, 95].

In summary, electroosmotic micropumps harness electroosmosis to enable precise fluid control in microfluidic systems. They offer advantages such as reliability, compatibility with delicate samples, and the ability to achieve smooth speed profiles. However, considerations such as the requirement for electrically conductive solvents and high voltage should be taken into account. Ongoing research aims to address these limitations and further enhance the performance of EO micropumps.

4.4.12 Bubble-Type Micropumps

Bubble-type micropumps operate by utilizing the intermittent collapse and expansion of bubbles in a microchannel under regulated voltage input to generate pumping action [5]. Figure 4.8a provides a schematic depiction of a collapsing and expanding bubble-type micropump. However, bubble-type micropumps typically require heating, which can limit their usage in situations where heating is not feasible or desirable. Lin and Tsai researched a valveless micropump that used thermal bubble stimulation and nozzle/diffuser stream control. The pumping chamber was formed by inducing the formation of a microbubble inside the microchamber. As the bubble became larger, the rate of flow at the diffuser (Qd) surpassed the rate of flow at the nozzle (Qn). Upon the bursting of the bubble, the quantity of Qd decreased relative to Qn. Achieving a unidirectional flow from the nozzle to the diffuser was made possible by controlling the voltage input during each bubble collapse and growth. The micropump construction, including channels, diffuser/nozzle flow regularizer, and the pumping chamber, was manufactured on a silicon substrate. The bubble-type micropump attained a peak flow rate of five liters per minute when the applied voltage was consistently administered at a frequency of 250 Hz, using 1 W of power and operating with a duty cycle of 10% [15]. Geng et al. showed a bubble-based micropump for moving fluids in another work. The device raised the water level by a few millimeters when the flow rate was set to 100 L per minute. But because this micropump used a high AC voltage, the area between the two channels got very hot. Zahn et al. created an on-chip MEMS bubble micropump that is combined with microneedles to make devices that deliver drugs continuously. The micropump used thermally produced bubbles that grew and shrunk with the help of flow-rectifying check valves to keep the flow rate under control. An SOI production method and quartz dice were used to make the gadget. The flow rate that was recorded was about 0.12 l/min, and the pressure was 3.9 kPa. The integration of microneedles with the micropump allowed precise control of the injection flow for specific drug concentrations in the drug delivery system [5].

Bubble-type micropumps can also be utilized in specimen collection. By reversing the direction of the valve, the flow direction can be reversed, enabling fluid extraction using a micropump. People with diabetes may be able to measure their blood glucose levels using this integrated gadget. A simple micropump that worked by periodically expanding and contracting a vapor bubble inside a microchannel was described by Yin and Prosperetti. Using laser machining, the micropump was created

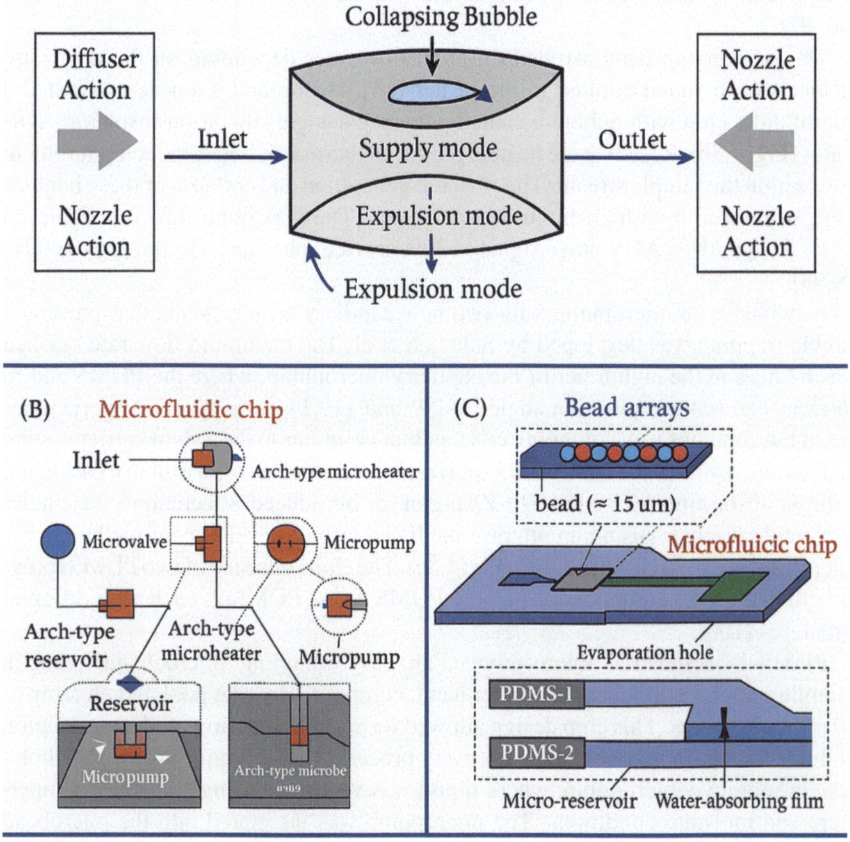

Fig. 4.8 **a** The schematic of Bubble-Type Micropump[15]. **b** An optical picture (lower) and a schematic view (upper) of a microfluidic chip with a micromixer, a microvalve, and a micro pump. **c** Diagram illustrating the functionality of a microscale pump propelled by the combined forces of evaporation and capillary phenomena. Reprinted with permission from Ref. [5]. Copyright 2018, Elsevier

on an acrylic substrate with a diameter of 150 μm. Pump rates of several tens of l/min were achieved with a channel diameter of 100 μm and pressure variations of several kPa [5]. When joule heating is utilized to generate the bubble, these micropumps are appropriate for pumping electrically conducting liquids, such as saline solutions, in biomedical applications. To heat nonconducting fluids, microchannel heaters must be employed. Furthermore, in situations requiring the fusion of multiple types of quantities via collapsing/expanding cycles, the utilization of a parallel micropump configuration in two microchannels linked at a Y-junction can prove advantageous. Jung and Kwak devised a micro pump that utilized bubbles and incorporated a microheater. By incorporating microheaters into a silicon dioxide substrate that served as the base plane of a silicon wafer, the micropump was manufactured. A pump chamber and valveless diffuser/nozzle components comprised its construction. A glass wafer comprised the intake and output apertures on the upper plate of the micropump [96, 97].

The micropump demonstrated different flow rates depending on the duty ratio of the voltage signal applied to the heater [15]. Huang and Tsou demonstrated a microfluidic chip with bubble-actuated pumping using a silicon-on-insulator (SOI) wafer (Fig. 4.8b). Resistive arc heaters produced thermal bubbles, inducing turbulent flow within the sample stream. The periodic generation and collapse of these bubbles were stimulated by a high-frequency AC pulse. The maximum flow rate achieved was 4.5 L/s with an AC voltage signal of 30% service speed and a frequency of 60 Hz [5, 98].

A bubble-type micropump with two new capillary arrangements that prevented bubble trapping was developed by Safavieh et al. The maximum flow rate reached was 62 nL/s in the eighth tier of the capillary micropump, where the PDMS and Si surfaces had static connection angles of 29° and 119.1°, respectively. Vaporization-based micropumps use a pumping concept that is similar to the xylem carrier system in trees and involves the controlled vaporization of a fluid across a film in a gas region with an absorption factor [5, 99]. Zhang et al. introduced a technique for single-nucleotide discoveryusing an apyrase-mediated primer development method and a microbead microfluidic array chip (Fig. 4.8c). The chip consisted of two PDMS (polydimethylsiloxane) components, namely PDMS-1 and PDMS-2, each with different structures [100].

PDMS-1 included a micro-groove, an easy sampling microchannel, and a sampling pool. PDMS-2, on the other hand, comprised a waste pool and an array of 10 microchambers. This chip design allowed for specific functions and manipulation of fluids for the single-nucleotide discovery process. The technique utilized a bubble-type evaporative micropump, where liquid was vaporized using controlled temperature and moisture conditions. The micropump was integrated into the microbead microfluidic array chip. By adjusting the vaporization temperature (ranging from 45°C to 20 °C), relative moisture (65%), and vaporization area (40 mm^2), the liquid flow velocity could be regulated within the range of 120–20 m/s. The chip's capability for single-nucleotide discovery was evaluated, and it demonstrated a detection limit of 0.2 pM for the target DNA. This suggests that the technique and chip design were capable of effectively detecting and analyzing specific nucleotide sequences [101].

4.4.13 Electro-Wetting Micropump

Electro-wetting micropumps are microfluidic devices that leverage the phenomenon of electro-wetting to manipulate and transfer fluids, In Fig. 4.2, the schematic is shown. Electro-wetting exploits the surface tension forces that exist at the microscale, allowing for precise control by applying a voltage potential between a dielectric-coated electrode and a conductive liquid. This voltage potential alters the interfacial energy between the solid and liquid surfaces, resulting in the movement of the liquid. The operational concept of an electro-wetting micropump is shown in figure. The motion of the liquid is initiated by reducing the interfacial energy by the application of voltage to the dielectric layer. Continuous electro-wetting (EW) is often used to manipulate the surface tensions of two immiscible liquids, such as liquid-phase metal (mercury) and electrolyte [15]. Electro-wetting is a reversible phenomenon, meaning that the system reverts to its initial condition when the voltage is no longer applied. An important benefit of electro-wetting is its capacity to control specific amounts of liquid, allowing for the creation of drops with precise volumes in small-scale systems. In the case of a continuous electro-wetting micropump, the actuation energy is derived from the surface tension-induced motion of a mercury drop within a microchannel filled with electrolyte. This micropump is typically composed of interconnected wafers. A glass wafer with a microchannel, filled with electrolyte, is utilized, while a second silicon wafer features a deflection membrane. The motion of the mercury droplet induces a pulling force on the electrolyte, resulting in the deflection of the membrane. Operating at a driving voltage of 2.3 V and consuming 170 W of power, this micropump can achieve a volumetric flow rate of up to seventy liters per minute. Additionally, when a voltage of 2.3 V is applied at a frequency of 25 Hz, the micropump can generate a maximum pressure of approximately 0.8 kPa [15, 31, 97].

In summary, electro-wetting micropumps harness the unique capabilities of electro-wetting in microfluidics. By precisely controlling surface tension through voltage manipulation, these micropumps offer the ability to manipulate and transfer fluids at the microscale. Their advantages include the generation of drops with specific volumes and their suitability for various microfluidic applications such as drug delivery systems, lab-on-a-chip devices, and micro-scale pumping and mixing operations.

4.4.14 Flexural Planar Wave (FPW) Micropumps

Flexural Planar Wave (FPW) micropumps are a type of micropump that utilize the phenomenon of flexural waves to generate fluid flow at the microscale (Fig. 4.9). These micropumps are designed to manipulate small volumes of liquid with precision and control, making them valuable tools in various applications such as microfluidics, biomedical engineering, and lab-on-a-chip systems. The operating principle of FPW micropumps involves the generation of flexural waves using piezoelectric actuators. Flexural waves, also known as bending waves, are mechanical waves that propagate through thin plates or membranes. In FPW micropumps, these waves are produced in a planar format, generating an acoustic field that interacts with the surrounding liquid. The flexural waves induce fluid motion through a phenomenon called acoustic streaming. Acoustic streaming occurs when the flexural waves transfer momentum to the liquid, resulting in controlled flow. This flow can be directed and manipulated by adjusting the characteristics of the flexural waves, such as their frequency, amplitude, and spatial distribution. FPW micropumps offer several advantages over other types of micropumps. Firstly, they operate at lower voltages and consume less power, making them energy-efficient and suitable for portable and battery-powered devices. Additionally, FPW micropumps do not require additional components such as heaters or valves, simplifying their design and reducing the risk of component failure. The design and fabrication of FPW micropumps involve careful consideration of materials, dimensions, and manufacturing techniques. Thin plates or membranes made of materials like silicon nitride or polymers are used to generate flexural waves. Piezoelectric actuators, typically based on materials like zinc oxide or piezoelectric ceramics, are integrated into the micropump structure to induce the flexural waves. For example, Moroney et al. conducted experiments on FPW pumping, observing fluid motion induced by ultrasonic flexural waves scattering in a thin membrane. They achieved fluid pumping at a flow velocity of up to 0.255 L per minute in silicone tubes. FPW micropumps have found applications in various fields. In drug delivery systems, they offer precise control over fluid flow rates and volumes, enabling targeted and controlled administration of medications. In microanalysis systems, FPW micropumps facilitate the manipulation and analysis of microscale samples, allowing for efficient and high-throughput experiments. They are also used in cell manipulation systems, where the precise handling and positioning of cells are crucial for research and medical purposes. Nguyen and coworkers developed a microfluidic system using the FPW micropump. Their micropump was constructed utilizing a conventional silicon nitride sheet with a width ranging from 3 to 1 μm, as well as low-stress aluminum, zinc oxide, and polysilicon. The 10 mm × 10 mm microfluidic system was produced using conventional manufacturing processes. The FPW micropump exhibited lower power consumption and voltage requirements (less than ten milliwatts), making it suitable for handling sensitive biomaterials. Potential applications include drug delivery systems, cell manipulation systems, and total microanalysis systems (TAS). Meng et al. proposed a micromachined micropump that creates an acoustic field in the liquid surrounding the membrane by use of an ultrasonic flexural

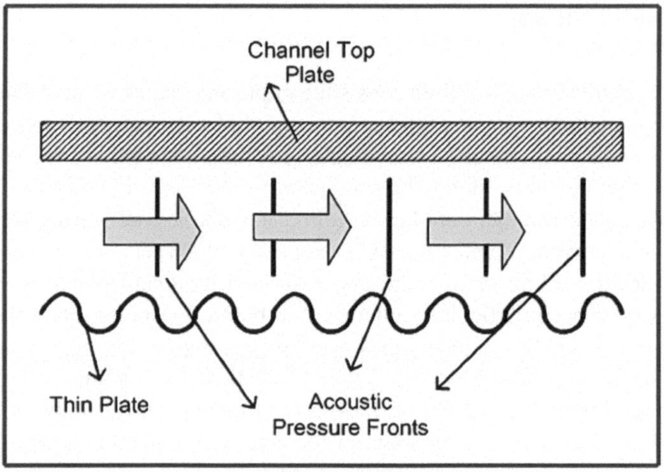

Fig. 4.9 Schematic illustration of acoustic streaming [15]

wave that travels down a thin membrane. The liquid flows because of this acoustic field. The micropump has a special set of radial transducers that allow for focussed and bidirectional flow. This type of micropump finds applications in micro total analysis systems and medication delivery devices [15, 102, 103].

4.4.15 Magnetohydrodynamic (MHD) Micropumps

Magnetohydrodynamic (MHD) micropumps are a type of microfluidic device that utilizes the principles of MHD to pump and manipulate fluids at the microscale (Fig. 4.2). MHD is the science of examining the magnetic characteristics of electrically conducting liquids. The working principle of MHD micropumps involves the interaction between a magnetic field, an electric current, and a conductive fluid. Typically, a microchannel or microfluidic channel contains embedded electrodes that generate the necessary electric current. A magnetic field is imposed orthogonally to the direction of fluid motion. The interaction of an electric current with a magnetic field generates Lorentz forces inside the fluid. The Lorentz force is generated by the interaction between an electric current and a magnetic field, which produces a force that drives the fluid in a certain direction, allowing for pumping and manipulation. These micropumps possess several benefits compared to conventional mechanical pumps, including their small size, low energy consumption, and lack of moving components, which decrease the likelihood of mechanical breakdown [104–106].

4.5 Blood Transport

Blood is a complex biological fluid consisting of plasma, platelets, white blood cells, and red blood cells. Plasma, the fluid component of blood, comprises approximately fifty-five percent of the total blood volume [107]. As the primary mode of transportation within the body, blood plays a crucial role in overall bodily function. The heart pumps blood through capillaries, veins, and arteries, delivering oxygen and nutrients while preventing infections. Consequently, blood serves as a vital indicator of human health, aiding in accurate diagnosis and treatment planning. Recently, researchers have been particularly interested in blood transportation in the context of blood tests and analytical techniques, with micropumps emerging as a frequently employed technology for blood transfusion [107, 108]. Micropumps play a pivotal role in facilitating controlled blood transport within the human body and medical devices. They offer a promising solution for various applications related to blood transport, including drug delivery, artificial organs, and lab-on-a-chip systems. One of the key challenges in blood transport is maintaining a controlled and consistent flow. Micropumps address this challenge by providing precise control over the flow rate and volume of blood. They can be seamlessly integrated into medical devices or implanted directly into the body to assist with blood circulation or deliver medications [109, 110].

In drug delivery applications, micropumps prove invaluable in delivering therapeutic agents directly into the bloodstream. By regulating the flow rate and timing of delivery, micropumps enable precise dosing and targeted administration of medications. This approach is particularly beneficial for conditions requiring continuous or controlled drug delivery, such as diabetes management or chemotherapy. Micropumps also find utility in artificial organs and assistive devices like artificial hearts or ventricular assist devices (VADs). These devices rely on micropumps to propel blood throughout the body, providing circulatory support for patients with heart failure or other cardiac conditions. The micropumps employed in such applications are designed to mimic the pumping action of the heart and maintain a consistent blood flow. Lab-on-a-chip systems, which amalgamate multiple laboratory functions onto a single chip, benefit from micropump technology for blood transport. These systems often necessitate precise manipulation of small blood samples for analysis and diagnostics. By incorporating micropumps into these devices, the movement of blood samples can be effectively controlled, facilitating efficient and automated testing processes. The design and implementation of micropumps for blood transport require careful consideration of biocompatibility, reliability, and power consumption. It is vital to employ biocompatible materials and fabrication techniques to ensure the micropumps do not elicit adverse reactions or cause damage to the blood or surrounding tissues. Reliability is of utmost importance to prevent failures or malfunctions that could have severe consequences for patients. Additionally, micropumps intended for implantation should exhibit low power consumption to minimize their impact on battery life and power requirements. Cardiac assist pumps are employed in vivo or clinical settings to provide circulatory support to failing hearts. A disposable pump is necessary for a blood cancer therapy device to attain a significant flow rate of 100 mL/min under a pressure head of two kilopascals. Taylor et al. [111] utilized microfluidic modules that integrated micropump technology to control and convey biological samples in preparation for blood processing involving genomic DNA [31, 112–114]. Micromed Cardiovascular, Inc.'s DeBakey VAD is presently enduring clinical trials and has been approved for experimental use on humans. By utilizing an implanted axial flow pump, a ventricular assist pump was developed to enhance blood flow within the heart. It can deliver continuous flow via the heart, a capability not achievable with existing implanted assist devices [107, 114, 115].

Ongoing research and development in the field of micropumps continue to explore novel technologies and techniques to enhance blood transport applications. These advancements primarily focus on improving pump efficiency, reducing size and power requirements, and ensuring long-term reliability. By harnessing the potential of micropumps, researchers and medical professionals aim to improve patient outcomes, expand treatment options, and advance the field of blood transport in various medical contexts.

4.6 Drug Delivery

There have been several reports on various micropumps used for medication delivery applications. In this section, we will explore several representative micropumps designed for medication delivery.

4.7 Drug Delivery for Epilepsy

50 million people worldwide suffer from epilepsy; many of them are not good candidates for surgery, have seizures that are not controlled by medicines, or are resistant to therapy. Whether a seizure leaves a lasting impression depends on where it occurs in the cerebral cortex. Recurrent seizures are still linked to high rates of morbidity and death even with the great progress made with antiepileptic drugs, conventional therapies have not been able to effectively treat them. Direct drug delivery systems (DDS) are a potential class of implanted devices that use micropumps to provide local medication delivery. Most micropumps that have been documented employ a motion diaphragm, such as a PDMS membrane that is deflected by heating or electromagnetic events. The integrated DDS and seizure detector comprise the two components of the suggested responsive DDS, as seen in Fig. 4.10A. A signal processor and preamplifier are part of the detector. Brain impulses are recorded using an intracortical electroencephalograph (EEG) collection device. The dedicated DDS is triggered when the seizure detector feeds input to the micropump system. Our findings indicate that the key components of a seizure detector are a demodulator, a low-pass filter,

and a chopper-stabilized preamplifier. It was tested on three drug-resistant individuals using EEG recordings (ages: 24–35), with each subject experiencing an average of five seizures. Two weeks of EEG recording were conducted to survey and monitor the patients. The primary focus of this research is to create a micropump and integrate a DDS prototype with a microelectronic control module. This paper lays the groundwork for creating an implantable version of the proposed embedded DDS by describing and presenting initial findings. During seizures, the electromagnetic micropump device can quickly deliver therapeutic fluid. This system comprises an electromagnetic micropump and a microelectronic control system [116–118].

The micropump has a single chamber and two passive valves—one for suction and one for pumping that allow fluid to flow in only one direction (Fig. 4.10B(a)). It utilizes a stress-deflectable PDMS membrane diaphragm. The deflection is caused by a pressure difference, resulting in volume fluctuations in the micropump chamber. Fluid enters and remains in the chamber with each deflection of the membrane (Fig. 4.10B(b)). A cylindrical magnetic moving component (MMP) connected to the membrane diaphragm is responsible for the deflection. When an electrical current (I) passes through a coil, an electromagnetic field (B) is generated, which presses on this cylindrical part (Fig. 4.10B(c)) [119].

4.8 Drug Delivery to the Mouse Inner Ear Using an Implanted Micropump with Nanoliter Resolution

Delivering drugs to the inner ear of mice requires a highly precise and controlled approach. According to a 2018 report by the World Health Organization, 466 million people worldwide suffer from hearing loss. However, current methods for inner ear drug delivery, such as injections and oral administration, have proven ineffective due to challenges such as the blood-cochlear barrier, unintended drug distribution, potential toxicity, and inconsistent drug concentration levels in the cochlea. The cochlea, responsible for auditory perception in the inner ear, presents significant obstacles to medication delivery due to its small size and limited accessibility. This makes it challenging to develop effective treatments for repairing or protecting damaged or missing cochlear cells in cases of deafness or hearing loss. To overcome these challenges, researchers have been exploring advanced microsystems and innovative techniques. Mouse models with specific genetic modifications, such as knock-in, knock-out, and transgenic variations, are crucial to study and develop treatments for hearing-related conditions. However, the small size of the mouse cochlea presents unique difficulties in delivering medications to the inner ear. While the human cochlea contains approximately 150–200 μL of perilymph, the mouse cochlea holds only about 620 nanoliters. This limited volume makes the cochlea particularly vulnerable to extended or high-speed intracochlear drug delivery. To address these challenges, researchers have been developing microsystems-based approaches for safe and efficient inner ear drug delivery. However, commercially available microsystem-based delivery systems have

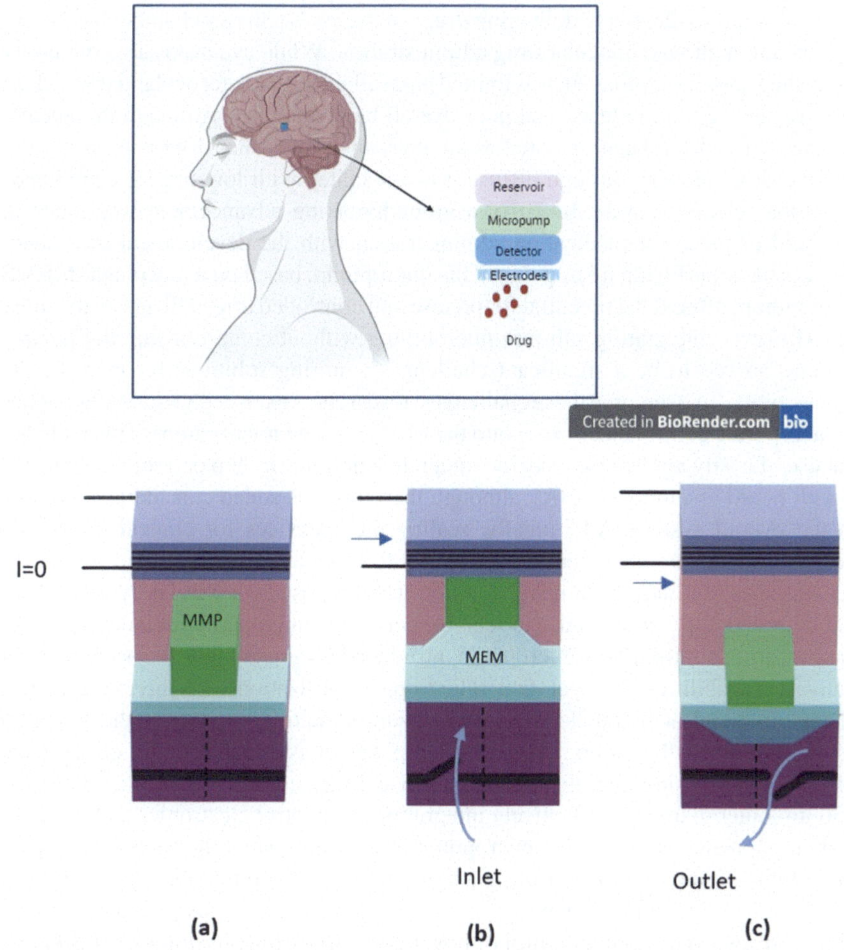

Fig. 4.10 **A** A depiction of a drug delivery system that can be implanted and embedded within the body. This system includes a seizure detector and a micro pump system. **B** Schematic of the proposed electromagnetic pump. (**a**) Close mode, (**b**) suction mode, and (**c**) pumping mode (MMP: Magnetic moving part, MEM: membrane, B: Magnetic field, I: Electrical current)

limitations. For example, existing osmotic pumps lack precision in drug delivery and cannot be stopped or started once the infusion has begun. They also do not allow for adjusting the flow rate during the delivery process. In recent experiments conducted with guinea pigs, a peristaltic micropump called iPrecio® SMP-300, combined with a microcatheter, was used to infuse a concentration marker into the cochlea. However, there is a need for a compact and lightweight pumping system that can provide ultra-low flow rates, precise time-sequential control over extended periods (potentially months), remote activation and deactivation, and programmable automated deliveries. Similarly, effectively delivering drugs to the eye's anterior chamber has been a significant challenge in ocular drug administration. While eye drops are commonly used, they have limitations such as limited bioavailability to target ocular tissues, short drug residence time in tears, and poor overall bioavailability. Although therapeutic contact lenses have been explored as an alternative for ocular drug administration, passive drug delivery through diffusion is associated with low patient compliance and long release periods. In 2016, a groundbreaking advancement was made in the field of inner ear medication administration with the development of a nanoliter cochlear perfusion micropump. This micropump, based on a tiny planar MEMS micropump, offered the potential for precise and controlled drug delivery to the inner ear. However, integrating sub-mm microtubing without compromising the pump's volume proved to be a significant challenge, requiring solutions for microfluidic connectivity. To overcome these challenges, researchers explored various approaches to incorporate control electronics into the MEMS-based micropumps. This included the use of costly application-specific integrated circuits on-chip or a separate printed circuit board assembly (PCBA), although these options added considerable volume to the overall system. Additionally, scaling these devices for clinical translation posed further difficulties, requiring changes to fabrication methods and masks. To address these limitations, a novel solution was proposed—a wirelessly controlled, miniaturized, fully implanted, and scalable peristaltic micropump for inner ear medication administration. This micropump introduced programmable and accurate drug delivery capabilities. Its operation relied on the utilization of a thermally phase-change material, which underwent crystallization/melting to generate the actuation force needed for the pump. The micropump was constructed directly on the back of a PCBA, encasing microtubing with precise dimensions (125 μm inner diameter, 250 μm outer diameter), effectively integrating the control electronics and mechanical components. Notably, this micropump design eliminated the need for complex microfluidic interconnects, ensuring a simplified and biocompatible flow route that was both leak-free and efficient. In vitro evaluations demonstrated the micropump's ability to achieve nanoliter control of flow rates, surpassing physiological backpressures by tenfold. Encouragingly, in vivo experiments conducted on a mouse model successfully delivered drugs through the round window membrane (RWM), a criti cal barrier in inner ear drug administration. Furthermore, extensive biocompatibility tests were performed to assess the suitability of the micropump components for subcutaneous implantation and potential clinical translation. These tests confirmed the safety and compatibility of the micropump, as a prototype was implanted for a month without any signs of infection or inflammation.

With its affordability and practical design, this micropump fabrication approach holds the potential for broader applications in larger animals and clinical settings. By adjusting the microtubing diameter and actuator volume, this innovative solution can be tailored to meet the specific requirements of diverse patient populations, including adults and children. In the field of ocular drug delivery, researchers have also been working on overcoming challenges associated with conventional methods. Devices using external stimuli such as ultrasound, laser light, and magnetic and electric fields have been developed for controlled drug delivery to the eye. However, these devices have their drawbacks, including the need for batteries, complex constructions, and the risk of exposing the treated retina to harmful substances. To address these issues, a proposed solution involves embedding a micropump within a contact lens. This micropump can be wirelessly operated using an external magnetic field, eliminating the need for batteries and complex constructions. It consists of various components, including an actuation membrane, a magnetic nanoparticle-polydimethylsiloxane (PDMS) composite for stimulation, a drug-filling chamber, and a micro-check valve for one-way drug administration. The strength and frequency of the magnetic field can be adjusted to control the dosage and timing of drug release. The micropump can be incorporated into soft or hard contact lenses, either secured with glue or integrated into stiff contact lenses using a photolithography-based mold [120, 121].

In summary, these innovative drug delivery systems offer promising solutions for targeted inner ear medication delivery in mice and controlled drug administration for ocular disorders. By overcoming the challenges associated with conventional drug delivery methods, these advancements pave the way for more effective treatments for hearing loss and ocular conditions in the future.

4.9 Conclusion

In recent years, significant advancements have been made in the development of micro-electro-mechanical systems (MEMS) for controlled drug release applications, particularly in the field of medicinal drug delivery systems. The ability to deliver pharmacological cargo with precise dosages has been a focus of research in this area. MEMS technology has played a pivotal role in the rapid growth of microfluidic devices, with micro-pumps being a vital component. These micro-pumps, capable of delivering flow rates in milliliters (ml) or microliters (μl) per minute, have been carefully designed and manufactured for effective in vivo administration of medicinal substances. They can be implanted or used in transdermal applications. In vitro studies have shown promising results in achieving extended and controlled release patterns of medicinal substances, such as chemotherapeutic drugs and hormones. However, further research is still required to fully explore their potential. Microfluidic fluid transfer, as a relatively new and appealing technology, has gained increased interest in biomedical applications. Various micropump architectures have been described, utilizing different materials, with silicon being a commonly used substrate material for microfluidic devices. Among the different types of micropumps, piezoelectrically actuated mechanical displacement micropumps have received considerable attention. However, the applied voltage remains a limiting factor in their drug delivery capabilities. Electrostatic and piezoelectric micropumps typically require high driving voltages for efficient operation. On the other hand, conducting polymer film actuators, such as ion conductive polymer film (ICPF) actuators, show promise as low-voltage micropumps, offering potential advantages in drug delivery systems.

Micropumps have demonstrated their potential in delivering medication to inaccessible or restricted areas, such as rapid drug delivery during seizures for epilepsy patients or delivering medications to the inner ear for the treatment of hearing problems. They have also been utilized as in vivo heart aid pumps for pumping blood in cases of heart failure. In conclusion, the advancements in MEMS technology and microfluidic devices, particularly in the field of micropumps, have opened up new possibilities for controlled drug delivery and biomedical applications. Further research and development are essential to optimize their performance, enhance drug release capabilities, and explore new avenues for their utilization in healthcare and therapeutic interventions. With continued innovation, micropumps hold great promise in improving patient outcomes and addressing unmet medical needs.

References

1. Chen, L., et al.: Millifluidics, microfluidics, and nanofluidics: manipulating fluids at varying length scales. Mater. Today Nano **16**, 100136 (2021)
2. Niculescu, A.-G., et al.: Fabrication and applications of microfluidic devices: a review. Int. J. Mol. Sci. **22** (2021). https://doi.org/10.3390/ijms22042011
3. Chappel, E., Dumont-Fillon, D.: Micropumps for drug delivery. Drug Deliv. Dev. Ther. Syst. 31–61 (2021)
4. Spencer, W. J., et al.: Some engineering aspects of insulin delivery systems. Diabetes Care, **3**(2), 345–350 (1980).
5. Wang, Y.-N., Fu, L.-M.: Micropumps and biomedical applications—a review. Microelectron. Eng. **195**, 121–138 (2018)
6. Mohith, S., Karanth, P.N., Kulkarni, S.M.: Recent trends in mechanical micropumps and their applications: a review. Mechatronics **60**, 34–55 (2019)
7. Ashraf, M.W., Tayyaba, S., Afzulpurkar, N.J.: Micro electromechanical systems (MEMS) based microfluidic devices for biomedical applications. Int. J. Mol. Sci. **12**(6), 3648–3704 (2011)
8. Villarruel Mendoza, L.A., et al.: Recent advances in micro-electro-mechanical devices for controlled drug release applications. Front. Bioeng. Biotechnol. **8**, 827 (2020)
9. Bußmann, A., et al.: Piezoelectric silicon micropump for drug delivery applications. Appl. Sci. **11** (2021). https://doi.org/10.3390/app11178008
10. Sahini, M.G., Banyikwa, A.T.: Chapter 25—electronic drug delivery systems. In: Nayak, A.K., et al. (eds.) Advanced and Modern Approaches for Drug Delivery, pp. 703–732. Academic Press (2023)
11. Rahmanian, M., et al.: A micropillar array-based microfluidic chip for label-free separation of circulating tumor cells: the best micropillar geometry? J. Adv. Res. **47**, 105–121 (2023)
12. Sartipzadeh, O., et al.: Microfluidic-assisted synthesis and modeling of stimuli-responsive monodispersed chitosan microgels for drug delivery applications. Sci. Rep. **12**(1), 8382 (2022)
13. Sartipzadeh, O., et al.: Controllable size and form of droplets in microfluidic-assisted devices: effects of channel geometry and fluid velocity on droplet size. Mater. Sci. Eng. C **109**, 110606 (2020)
14. Sartipzadeh, O., et al.: Microfluidic-assisted synthesis and modelling of monodispersed magnetic nanocomposites for biomedical applications. Nanotechnol. Rev. **9**(1), 1397–1407 (2020)
15. Nisar, A., et al.: MEMS-based micropumps in drug delivery and biomedical applications. Sens. Actuators B Chem. **130**(2), 917–942 (2008)

16. Junwu, K., et al.: Design and test of a high-performance piezoelectric micropump for drug delivery. Physical **121**(1), 156–161 (2005)
17. Yaghoubi, F., et al.: A functionalized graphene oxide with improved cytocompatibility for stimuli-responsive co-delivery of curcumin and doxorubicin in cancer treatment. Sci. Rep. **12**(1), 1959 (2022)
18. Yaghoubi, F., et al.: Multiresponsive carboxylated graphene oxide-grafted aptamer as a multifunctional nanocarrier for targeted delivery of chemotherapeutics and bioactive compounds in cancer therapy. Nanotechnol. Rev. **10**(1), 1838–1852 (2021)
19. Shahidi, M., et al.: HB5 aptamer-tagged graphene oxide for co-delivery of doxorubicin and silibinin, and highly effective combination therapy in breast cancer. Cancer Nanotechnol. **14**(1), 59 (2023)
20. Mazidi, Z., et al.: Smart stimuli-responsive implantable drug delivery systems for programmed and on-demand cancer treatment: an overview on the emerging materials. Chem. Eng. J. **433**, 134569 (2022)
21. Abtahi, N.A., et al.: Multifunctional stimuli-responsive niosomal nanoparticles for co-delivery and co-administration of gene and bioactive compound: In vitro and in vivo studies. Chem. Eng. J. **429**, 132090 (2022)
22. Abtahi, N.A., et al.: Development of highly efficient niosomal systems for co-delivery of drugs and genes to treat breast cancer in vitro and in vivo. Cancer Nanotechnol. **13**(1), 1–18 (2022)
23. Abtahi, N.A., et al.: A comparative study on biopharmaceutical function of curcumin and miR-34a by multistimuli-responsive nanoniosome carrier: in-vitro and in-vivo. Front. Mol. Biosci. **9**, 1043277 (2022)
24. Abtahi, N.A., et al.: Smart stimuli-responsive biofunctionalized niosomal nanocarriers for programmed release of bioactive compounds into cancer cells in vitro and in vivo. Nanotechnol. Rev. **10**(1), 1895–1911 (2021)
25. Abtahi, N.A., et al.: Multi-sensitive functionalized niosomal nanocarriers for controllable gene delivery in vitro and in vivo. Cancer Nanotechnol. **14**(1), 1–18 (2023)
26. Chappel, E.: Chapter 7—implantable drug delivery devices. In: Chappel, E. (ed.) Drug Delivery Devices and Therapeutic Systems, pp. 129–156. Academic Press (2021)
27. Das, P.K., Hasan, A.B.M.T.: Mechanical micropumps and their applications: a review. AIP Conf. Proc. **1851**(1), 020110 (2017)
28. Rajabi, F., Bakhshi, A., Kazemi, G.: Drug delivery applications of mechanical micropumps. In: Journal Conference: International Conference on Applied Researches in Science & Engineering. Amsterdam, Netherlands (2021)
29. Amirouche, F., Zhou, Y., Johnson, T.: Current micropump technologies and their biomedical applications. Microsyst. Technol. **15**(5), 647–666 (2009)
30. Ashraf, M.W., Tayyaba, S., Afzulpurkar, N.: Micro electromechanical systems (MEMS) based microfluidic devices for biomedical applications. Int. J. Mol. Sci. 12, 3648–3704 (2011). https://doi.org/10.3390/ijms12063648
31. Barua, R., et al.: Chapter 14—advances in MEMS micropumps and their emerging drug delivery and biomedical applications. In: Nayak, A.K., et al. (eds.) Advances and Challenges in Pharmaceutical Technology, pp. 411–452. Academic Press (2021)
32. Woias, P.: Micropumps—past, progress and future prospects. Sens. Actuators B Chem. **105**(1), 28–38 (2005)
33. Prithvi, J., et al.: Critical review and exploration on micro-pumps for microfluidic delivery. In: Guha, K., et al. (eds.) MEMS and Microfluidics in Healthcare: Devices and Applications Perspectives, pp. 65–100. Springer Nature Singapore, Singapore (2023)
34. Kalra, S., un Nabi, M.: Implantable bio-MEMS applications: a review. In: 2017 Recent Developments in Control, Automation & Power Engineering (RDCAPE) (2017)
35. Joshitha, C., Sreeja, B., Radha, S.: A review on micropumps for drug delivery system. In: 2017 International Conference on Wireless Communications, Signal Processing and Networking (WiSPNET). IEEE (2017)

36. Huo, C., Bai, C., Zhang, P.: Micropumps for microfluidic devices and BioMEMS. J. Phys. Conf. Ser. IOP Publishing (2020)
37. Forouzandeh, F., et al.: A review of peristaltic micropumps. Sens. Actuators A **326**, 112602 (2021)
38. Chircov, C., Grumezescu, A.M.: Microelectromechanical systems (MEMS) for biomedical applications. J. Phys. Conf. Ser. **13**(2), 164 (2022)
39. Kim, H., et al.: An integrated electrostatic peristaltic 18-stage gas micropump with active microvalves. J. Microelectromech. Syst. **24**(1), 192–206 (2015)
40. Mohd Ghazali, F.A., et al.: Soft dielectric elastomer actuator micropump. Sens. Actuators A **263**, 276–284 (2017)
41. Nam, D. N. C., et al.: Design of an IPMC diaphragm for micropump application. Sens. Actuat. A: Physical, **187**, 174–182 (2012).
42. Santos, J., Lopes, B., Branco, P.J.C.: Ionic polymer–metal composite material as a diaphragm for micropump devices. Sens. Actuators A **161**(1), 225–233 (2010)
43. Asadi Dereshgi, H., Dal, H., Yildiz, M.Z.: Piezoelectric micropumps: state of the art review. Microsyst. Technol. **27**(12), 4127–4155 (2021)
44. Wang, X., et al.: Knudsen pumps: a review. Microsyst. Nanoeng. **6**(1), 26 (2020)
45. Van Lintel, H. T. G., et al.: A piezoelectric micropump based on micromachining of silicon. Sens. Actuat., **15**(2), 153–167 (1988)
46. Ma, H.-K., et al.: Development of a piezoelectric micropump with novel separable design for medical applications. Sens. Actuators A Phys. **236**, 57–66 (2015)
47. Ghazali, F.A.M., et al.: MEMS actuators for biomedical applications: a review. J. Micromech. Microeng. **30**(7), 073001 (2020)
48. Zhao, P., Cao, Y., Li, J.: Nonlinear electromechanical coupling behavior of 1–3 piezoelectric composites. Acta Mater. **59**(14), 5534–5543 (2011)
49. Trindade, M.A., Benjeddou, A.: Effective electromechanical coupling coefficients of piezoelectric adaptive structures: critical evaluation and optimization. Mech. Adv. Mater. Struct. **16**(3), 210–223 (2009)
50. Zhang, Y., Lee, C.-H.: Piezoelectric energy harvesting pedal integrated with a compliant load amplifier. Adv. Mech. Eng. **11**(1), 1687814018820142 (2019)
51. Song, W.H., Lichtenberg, J.: Thermo-pneumatic, single-stroke micropump. J. Micromech. Microeng. **15**(8), 1425 (2005)
52. Chee, P.S., et al.: Wireless powered thermo-pneumatic micropump using frequency-controlled heater. Sens. Actuators A **233**, 1–8 (2015)
53. Chia, B.T., Liao, H.-H., Yang, Y.-J.: A novel thermo-pneumatic peristaltic micropump with low temperature elevation on working fluid. Sens. Actuators A **165**(1), 86–93 (2011)
54. Chee, P.S., et al.: Thermal analysis of wirelessly powered thermo-pneumatic micropump based on planar LC circuit. J. Mech. Sci. Technol. **30**(6), 2659–2665 (2016)
55. Abi-Samra, K., et al.: Thermo-pneumatic pumping in centrifugal microfluidic platforms. Microfluid. Nanofluid. **11**(5), 643–652 (2011)
56. Hamid, N.A., et al.: Fabrication of thermo-pneumatic driven microactuator for fluid transport applications. Adv. Sci. Lett. **19**(10), 2854–2859 (2013)
57. Spieth, S., et al.: Approaches for drug delivery with intracortical probes. Biomed. Eng. Biomed. Tech. **59**(4), 291–303 (2014)
58. Bonk, S.M., et al.: Design and characterization of a sensorized microfluidic cell-culture system with electro-thermal micro-pumps and sensors for cell adhesion, oxygen, and pH on a glass chip. Biosensors **5**, 513–536 (2015). https://doi.org/10.3390/bios5030513
59. Jeroish, Z.E., et al.: Microheater: material, design, fabrication, temperature control, and applications—a role in COVID-19. Biomed. Microdev. **24**(1), 3 (2021)
60. Jebrail, M.J., et al.: World-to-digital-microfluidic interface enabling extraction and purification of RNA from human whole blood. Anal. Chem. **86**(8), 3856–3862 (2014)
61. Feeny, R.M., Puissant, N.L., Henry, C.S.: Degassed PDMS pump for controlled extraction from dried filter samples in microfluidic devices. Anal. Methods **8**(47), 8266–8271 (2016)

62. Zehnle, S., et al.: Centrifugo-dynamic inward pumping of liquids on a centrifugal microfluidic platform. Lab Chip **12**(24), 5142–5145 (2012)
63. Schwemmer, F., et al.: A microfluidic timer for timed valving and pumping in centrifugal microfluidics. Lab Chip **15**(6), 1545–1553 (2015)
64. Song, Y., et al.: Size-based cell sorting with a resistive pulse sensor and an electromagnetic pump in a microfluidic chip. Electrophoresis **36**(3), 398–404 (2015)
65. Kant, R., et al.: Optimization of design and characterization of a novel micro-pumping system with peristaltic motion. Microsyst. Technol. **19**, 563–575 (2013)
66. Dehghan Manshadi, M.K., et al.: Electroosmotic micropump for lab-on-a-chip biomedical applications. Int. J. Numer. Model. Electron. Netw. Dev. Fields **29**(5), 845–858 (2016)
67. Kumar, N., et al.: Development of a solenoid actuated planar valveless micropump with single and multiple inlet–outlet arrangements. J. Micromech. Microeng. **26**(7), 075013 (2016)
68. Kumar, R., Sharma, G., Kumar, M.: Size and temperature effect on thermal expansion coefficient and lattice parameter of nanomaterials. Mod. Phys. Lett. B **27**(25), 1350180 (2013)
69. Mohd Jani, J., et al.: A review of shape memory alloy research, applications and opportunities. Mater. Des. (1980–2015) **56**, 1078–1113 (2014)
70. Khandelwal, A., Buravalla, V.: Models for shape memory alloy behavior: an overview of modeling approaches. Int. J. Struct. Chang. Solids **1**(1), 111–148 (2009)
71. Holman, H., Kavarana, M.N., Rajab, T.K.: Smart materials in cardiovascular implants: shape memory alloys and shape memory polymers. Artif. Organs **45**(5), 454–463 (2021)
72. Singh Rajput, G., et al.: Areas of recent developments for shape memory alloy: a review. Mater. Today Proc. **62**, 7194–7198 (2022)
73. Shuxiang, G., Fukuda, T.: SMA actuator-based novel type of micropump for biomedical application. In: IEEE International Conference on Robotics and Automation, 2004. Proceedings. ICRA'04 (2004)
74. Büttgenbach, S.: Electromagnetic micromotors—design, fabrication and applications. Micromachines **5**, 929–942 (2014). https://doi.org/10.3390/mi5040929
75. Abhari, F., Jaafar, H., Yunus, N.A.M.: A comprehensive study of micropumps technologies. Int. J. Electrochem. Sci. **7**(10), 9765–9780 (2012)
76. Long, E., Stuart, D.: Effective dynamics for solitons in the nonlinear Klein–Gordon–Maxwell system and the Lorentz force law. Rev. Math. Phys. **21**(04), 459–510 (2009)
77. Redžić, D.V.: Einstein at a crossroads: the Lorentz force and time dependence of a charged particle mass. In: 2019 PhotonIcs & Electromagnetics Research Symposium—Spring (PIERS-Spring) (2019)
78. Yunas, J., et al.: Polymer-based MEMS electromagnetic actuator for biomedical application: a review. Polymers **12**, 1184 (2020). https://doi.org/10.3390/polym12051184
79. Pirmoradi, F.N., et al.: A magnetically controlled MEMS device for drug delivery: design, fabrication, and testing. Lab Chip **11**(18), 3072–3080 (2011)
80. Tsai, N.-C., Sue, C.-Y.: Fabrication and analysis of ionic conductive polymer films for micro drug delivery systems. Sens. Actuators A **141**(2), 670–676 (2008)
81. Attiguppe, A.P., Chatterjee, D., DasGupta, A.: A novel integrated transdermal drug delivery system with micropump and microneedle made from polymers. Micromachines **14**, 71 (2023). https://doi.org/10.3390/mi14010071
82. Bohm, S., et al.: Chip-integrated non-mechanical microfluidic pump driven by electrowetting on dielectrics. Lab Chip **24**, 2893–2905 (2024)
83. You, R., Fu, X., Duan, X.: Acoustofluidic based wireless micropump for portable drug delivery applications. In: 2021 43rd Annual International Conference of the IEEE Engineering in Medicine & Biology Society (EMBC) (2021)
84. Uvarov, I.V., et al.: A peristaltic micropump based on the fast electrochemical actuator: design, fabrication, and preliminary testing. Actuators **10**, 62 (2021). https://doi.org/10.3390/act10030062
85. Liu, Y., et al.: A wearable, minimally-invasive, fully electrochemically-controlled feedback minisystem for diabetes management. Lab Chip **23**(3), 421–436 (2023)

86. Uvarov, I.V., et al.: Fast electrochemical micropump for portable drug delivery module. Russ. Microlectron. **52**(3), 186–194 (2023)
87. Ma, C., Suzuki, H.: Microfluidic devices using electrochemical micropumps and microvalves for autonomous solution processing. ECS Meet. Abs. **MA2020-02**(66), 3378 (2020)
88. Tavari, T., et al.: A systematic overview of electrode configuration in electric-driven micropumps. Electrophoresis **43**(13–14), 1476–1520 (2022)
89. Liu, X., et al.: Electrochemical detection of ascorbic acid in finger-actuated microfluidic chip. Micromachines **13**, 1479 (2022). https://doi.org/10.3390/mi13091479
90. Villarruel Mendoza, L.A., et al.: Recent advances in micro-electro-mechanical devices for controlled drug release applications. Front Bioeng Biotechnol **8**, 827 (2020)
91. Peng, Y., et al.: A review on electrohydrodynamic (EHD) pump. Micromachines **14**, 321 (2023). https://doi.org/10.3390/mi14020321
92. Li, L., et al.: Advancement of electroosmotic pump in microflow analysis: a review. Anal. Chim. Acta **1060**, 1–16 (2019)
93. Wang, X., et al.: Electroosmotic pumps and their applications in microfluidic systems. Microfluid. Nanofluid. **6**(2), 145–162 (2009)
94. Wang, X., et al.: Electroosmotic pumps for microflow analysis. TrAC Trends Anal. Chem. **28**(1), 64–74 (2009)
95. Hossan, M.R., et al.: Review: electric field driven pumping in microfluidic device. Electrophoresis **39**(5–6), 702–731 (2018)
96. Tsai, N.-C., Sue, C.-Y.: Review of MEMS-based drug delivery and dosing systems. Sens. Actuators A **134**(2), 555–564 (2007)
97. Joshitha, C., Sreeja, B.S., Radha, S.: A review on micropumps for drug delivery system. In: 2017 International Conference on Wireless Communications, Signal Processing and Networking (WiSPNET) (2017)
98. Huang, C., Tsou, C.: The implementation of a thermal bubble actuated microfluidic chip with microvalve, micropump and micromixer. Sens. Actuators A **210**, 147–156 (2014)
99. Safavieh, R., Tamayol, A., Juncker, D.: Serpentine and leading-edge capillary pumps for microfluidic capillary systems. Microfluid. Nanofluid. **18**(3), 357–366 (2015)
100. Zhang, K., et al.: Sequencing genomes from single cells by polymerase cloning. Nat. Biotechnol. **24**(6), 680–686 (2006)
101. Gencturk, E., Mutlu, S., Ulgen, K.O.: Advances in microfluidic devices made from thermoplastics used in cell biology and analyses. Biomicrofluidics **11**(5), 051502 (2017)
102. Giri Nandagopal, M.S., et al.: Overview of microneedle system: a third generation transdermal drug delivery approach. Microsyst. Technol. **20**(7), 1249–1272 (2014)
103. Paul, S.R., et al.: MEMS-based controlled drug delivery systems: a short review. Polym. Plast. Technol. Eng. **55**(9), 965–975 (2016)
104. Dereshgi, H.A.: Investigation of electro-mechanical factors effecting micro-pump characteristics for biomedical applications. Sakarya Universitesi (Turkey) (2019)
105. Cardoso, R.M., et al.: A multi-pump magnetohydrodynamics lab-on-a-chip device for automated flow control and analyte delivery. Sensors **20**, 4909 (2020). https://doi.org/10.3390/s20174909
106. Seo, J.-H., et al.: Numerical investigations on magnetohydrodynamic pump based microchannel cooling system for heat dissipating element. Symmetry **12**, 1713 (2020). https://doi.org/10.3390/sym12101713
107. Amirouche, F., Zhou, Y., Johnson, T.J.: Current micropump technologies and their biomedical applications. Microsyst. Technol. **15**, 647–666 (2009)
108. Amirsahami, A., Barzinpour, F., Pishvaee, M.S.: A hierarchical model for strategic and operational planning in blood transportation with drones. PLoS ONE **18**(9), e0291352 (2023)
109. Gao, Q., et al.: Cell nucleus as endogenous biological micropump. Biosens. Bioelectron. **182**, 113166 (2021)
110. Vo, T.N.A., et al.: Engineering an extremely high flow rate micropump and integrating with an inertial microfluidics for rapid and efficient blood plasma extraction from fingertip blood with lancets. Sens. Actuators A **358**, 114430 (2023)

111. Taylor, M. T., et al.: Simulation of microfluidic pumping in a genomic DNA blood-processing cassette. J. Micromech. Microeng .**13**(2), 201 (2003)
112. Luo, X., Yang, L., Cui, Y.: Micropumps: mechanisms, fabrication, and biomedical applications. Sens. Actuators A **363**, 114732 (2023)
113. Tan, M., et al.: Recent advances in intelligent wearable medical devices integrating biosensing and drug delivery. Adv. Mater. **34**(27), 2108491 (2022)
114. Zheng, M., et al.: Microneedle biomedical devices. Nat. Rev. Bioeng. **2**(4), 324–342 (2024)
115. Palazzolo, T., et al.: Technology landscape of pediatric mechanical circulatory support devices: a systematic review 2010–2021. Artif. Organs **46**(8), 1475–1490 (2022)
116. Sayeed, M.A., et al.: iDDS: an edge-device in IoMT for automatic seizure control using on-time drug delivery. In: 2020 IEEE International Conference on Consumer Electronics (ICCE) (2020)
117. Zhao, X., et al.: Miniaturized neural implants for localized and controllable drug delivery in the brain. J. Mater. Chem. B **11**(27), 6249–6264 (2023)
118. Mariello, M., Eş, I., Proctor, C.M.: Soft and flexible bioelectronic micro-systems for electronically controlled drug delivery. Adv. Healthc. Mater. n/a(n/a), 2302969 (2023)
119. Hamie, A., Ghafar-Zadeh, E., Sawan, M.: An implantable micropump prototype for focal drug delivery. In: 2013 IEEE International Symposium on Medical Measurements and Applications (MeMeA). IEEE (2013)
120. Forouzandeh, F., et al.: A nanoliter resolution implantable micropump for murine inner ear drug delivery. J. Controll. Release **298**, 27–37 (2019)
121. Wang, C., Park, J.J.M., Letters, N.S.: Magnetic micropump embedded in contact lens for on-demand drug delivery. Micro Nano Syst. Lett. **8**(1), 1 (2020)

Chapter 5
Integration of Biosensors and Microfluidics and Their Application

Abstract A biosensor is a small analytical instrument or unit that measures one or more analytes by including a biological or biologically derived sensitive recognition element mounted on a physicochemical transducer. On the other hand, microfluidic systems have several advantages over traditional systems. They can reduce the sample and reagent volumes down to nanoliters, which is beneficial for reducing cost and waste. Microfluidic systems can also increase the mixing rate of different reagents, which improves the efficiency of chemical reactions. They allow for high-throughput processing, which means that a large number of samples can be processed quickly. Additionally, microfluidic systems improve transport for controlling flow conditions, which is important for achieving accurate and reproducible results. They also improve detection sensitivity, which helps detect small amounts of analytes. Another advantage of microfluidic systems is that they can be used for both sample preparation and detection, which simplifies the overall process. Because of these benefits, technologies allow biological and chemical components to be combined into a single platform, opening up new possibilities for future biosensing applications such as unprecedented accuracies, real-time detection, simultaneous analysis of multiple analytes in a single device, and portability. The purpose of this study is to highlight recent developments and achievements in microfluidic-based biosensing. The chapter presents examples from published works to illustrate the benefits of combining microfluidic and biosensing technologies, and the flexibility this approach provides for future biosensing applications. Diagnostic microsystems have simplified the process of diagnosing a condition by allowing doctors to pinpoint the exact location of the affected area and track disease progression over time. By using biosensors for multiplexed detection, the analysis throughput can be significantly increased, leading to a greater amount of data acquired in a single test. Microfluidic chips, which are a type of micro-total analysis system (μTAS), have proven to be an essential platform for high-throughput biosensors and portable. While microfluidic chips and biosensors are both powerful tools on their own, their combined capacity is particularly useful for analytical and diagnostic purposes.

Keywords Microfluidics · Biochip · Biosensor · Analytical instrument · Diagnostic microsystem · Real-time detection

S. M. Naghib and Y. Rezaian, *Frontiers of Nanobiotechnology*, Series in BioEngineering, https://doi.org/10.1007/978-3-031-88419-1_5

5.1 Introduction

In the last decade, biotechnology microsystems have gained popularity in laboratory-on-a-chip or microelectromechanical systems for processing, analyzing, and detecting biological pathogens and molecules [1]. These microdevices have several advantages, including reducing analysis costs, faster sample analysis times, multiplexing, and reduced quantities of reactants, as well as a lower risk of handling hazardous chemicals and pathogens [2, 3]. Biotechnology microdevices also have the potential to integrate with other analytical techniques and create low-cost, portable solutions, making them more sensitive than other microdevice techniques [4]. Biosensors are analytical tools used for medical diagnosis, agricultural and food safety, and environmental monitoring [5–9]. They consist of a biologically sensitive detection element attached to a detector to detect specific analytes in a sample. Biosensor selectivity is determined by the interaction between the target analyte and the biological detection element, generating signals that can be converted into readable electrical and optical signals by a transducer [10–13].

Microfluidic technology deals with the behavior, precise control, and manipulation of fluids on a scale below millimeters. It offers several advantages, including small size, manipulation of fluids and precise control, and the ability to simplify system operation and development. The unique behavior of fluids at the microscale distinguishes microfluidics from typical benchtop analytical, chemical, and biological equipment, as fluids travel almost exclusively inside the laminar flow regime due to intrinsically lower Reynolds numbers. This feature enables precise prediction and modeling of the flow profile [14–17]. Due to the small characteristic length of microfluidic devices, gravitational forces are insignificant, but surface and interfacial surface forces are prominent. The elevated proportion of surface area to volume grants the capability of microfluidic equipment to be conveniently transportable and advantageous for conducting examinations in the field. Capillary forces are another benefit of microfluidics, allowing for the manipulation of fluids in small spaces and the counteracting of gravitational forces. Capillary forces can also eliminate the need for pumps, whether integrated or not, simplifying system operation [18–21]. Microfluidics is classified into three types: droplet-based, continuous-flow, and digital microfluidics. Droplet-based microfluidic systems work by forming droplets in microchannels with two or more fluids that do not mix, typically at a T-junction. Permanently carved microchannels, coupled with peripheral devices like microvalves and micropumps, are utilized to control fluid flow in continuous microfluidic devices. Digital microfluidic systems are distinct from continuous microfluidic devices in that they offer precise control over individual droplets, which can be moved and manipulated on a planar array of electrodes through electrostatic actuation [22–24]. The implementation of microfluidics and lab-on-a-chip (LOC) contraptions for carrying out laboratory processes on a small scale is an immensely encouraging technique in the realm of biosensing. Portable microfluidic biosensors present unparalleled analytical performance, facilitating high-capacity testing and prompt detection at swift reaction rates. By virtue of their portability and sophisticated analytical abilities, these gadgets can be exploited in point-of-care (POC) techniques [25].

Lately, there have been noteworthy endeavors to merge biosensing and microfluidics, which enable users to take advantage of the small sample sizes, ease of integration and multiplexing, short turnaround times, and high mobility offered by microfluidics. Consequently, microfluidic chips have become one of the most remarkable technologies paired with an expanding range of biosensor systems, aimed at enhancing the general performance of the system [26].

5.2 Biosensors

Biosensors are frequently defined as a three-part system consisting of a signal-processing unit, a transducer, and a bioreceptor. A quantifiable signal is generated when the analyte interacts with the bioreceptor [6, 27–32]. The invention of enzyme electrodes in 1962 by Leland C. Clark initiated the development of biosensors. Since then, researchers from various fields, including chemistry, physics, VLSI, and material science, have collaborated to produce more advanced, trustworthy, and developed biosensing systems. Biosensors can have various definitions and terminologies depending on their application field [33]. Biosensors integrate biological materials and transducers to detect samples such as microbial load, pollutants, pharmaceuticals, metabolites, control parameters, and others. These devices translate biochemical signals into quantifiable physiochemical signals, which then quantify the quantity of the sample [10, 13, 34–43].

S. P. J. Higson and D. M. Fraser have presented two comprehensive explanations of a biosensor. As per the first definition, a biosensor is a chemical sensing instrument that merges a biologically derived recognition entity with a transducer. This blend enables the device to gauge a multifaceted biochemical parameter accurately. The other definition describes a biosensor as an analytical tool that unites a specific biological component with a physical element to transmit a detection occurrence. The term "biosensor" signifies the amalgamation of these two components. Biosensors can be classified based on the nature of their biological component or their transducer type (Fig. 5.1). The four primary kinds of transducers are electrochemical sensors, optical detection biosensors, thermal detection biosensors, and piezoelectric biosensors. Within the category of electrochemical biosensors, there are three subtypes: potentiometric, amperometric, and conductimetric [44–46].

Various biological elements can be paired with different transducers to create a diverse range of biosensors. Electrochemical biosensors are the most common type, accounting for over 80% of all biosensors. Depending on the specifics of their biological component, they may also be labeled as DNA sensors, immuno-sensors, or enzyme electrodes. Biosensor recognition elements are biological organisms that can identify a single substance, among many others. Nucleic acids, antibodies, enzymes, and receptors can fulfill this role [47]. The fundamental concept behind electrochemical biosensors lies in the interaction between the fixed biomolecule and the analyte

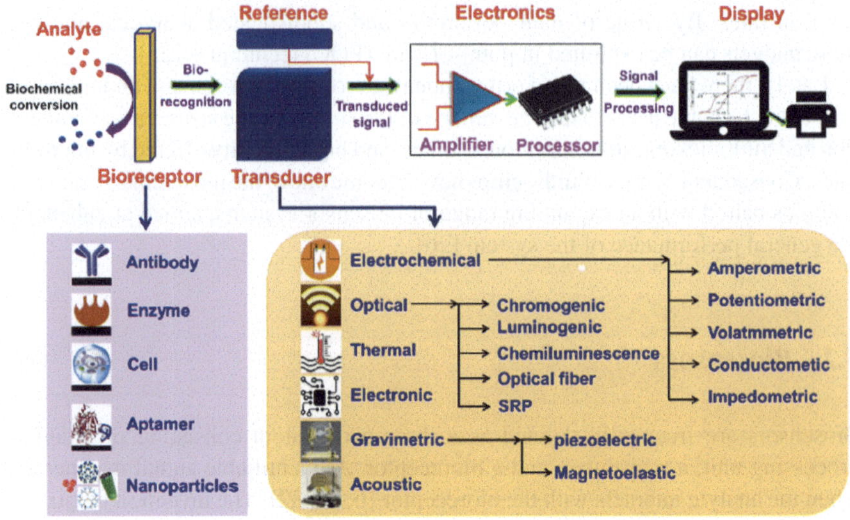

Fig. 5.1 A diagram illustrating the standard configuration of a biosensor, which includes a bioreceptor, transducer, electronic system (amplifier and processor), display (PC or printer), as well as different kinds of bioreceptors and transducers employed in biosensors, is also presented [46]

being targeted, which leads to the creation or utilization of electrons or ions. This chemical reaction has a significant impact on the electrical characteristics of the solution, including measurable features like potential or electric current [10].

5.3 Classification of Biosensors

5.3.1 Immunosensors

Antibodies are commonly used as bioreceptor in immuno-sensors to detect a specific antigen. Immunoassays are known for their precision, with the ability to detect a vast array of chemicals at very low levels. These tests are especially useful for detecting and quantifying proteins [48]. The majority of immuno-sensors are constructed using the solid-phase immunoassay concept, where the antigens and antibodies are immobilized on a solid support. This results in antigen–antibody interaction at the solid–liquid interface. The main benefits of immuno-sensors are their high selectivity and sensitivity, as well as the potential to enhance their affinity and selectivity by designing new recombinant antibodies [49]. The word immunoassay refers to tests based on immunoreactions, whereas the phrase immuno-sensor refers to whole instruments, i.e., immunoreactions-based biosensors. Figure 5.2 depicts a schematic illustration of an antibody's Y-shaped structure. Antibodies, which go by the scientific name of immunoglobulins, are great spherical plasma proteins that weigh in at a hefty 150 kDa. These proteins are commonly known as glycoproteins owing to their composition of two light and two heavy chains, which form the recognizable Y shape as shown in the Figure. They are produced in organisms in response to foreign substances, which are called antigens. Antibodies have a high binding affinity to target antigens, which allows them to detect analytes even in the presence of interfering molecules. Antibodies have a high binding affinity to target antigens, which allows them to detect analytes even in the presence of interfering molecules. Monoclonal and polyclonal antibodies are used in the construction of immuno-sensors [48].

Antibodies are proteins produced by the immune system in response to foreign substances, such as viruses and bacteria. They have the ability to recognize and bind to specific antigens with high affinity, making them useful in various applications, including disease diagnosis and treatment. Monoclonal and polyclonal antibodies are used in the construction of immuno-sensors. Polyclonal antibodies are commonly used in immunology research and diagnostic tests due to their ability to detect a wide range of antigens with high sensitivity. They are produced by different immune cells, each of which recognizes a specific epitope on the antigen. However, the relatively lower specificity of polyclonal antibodies can result in cross-reactivity, leading to false-positive results. This is because polyclonal antibodies can recognize different epitopes on the target antigen, as opposed to monoclonal antibodies that are produced by a single immune cell and bind to a single epitope on the antigen with high specificity. Despite their limitations, polyclonal antibodies remain valuable tools in immunology research and diagnostic tests. They are often used in the initial screening of potential antigens due to their high sensitivity. Furthermore, polyclonal antibodies can be used in conjunction with monoclonal antibodies to improve the

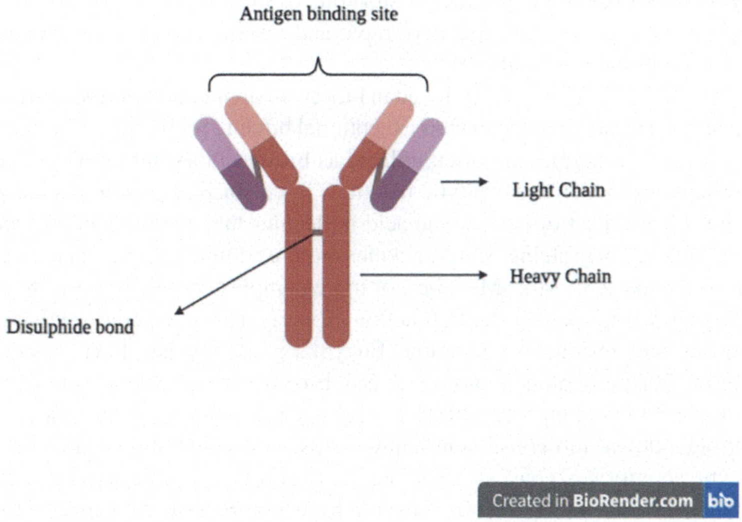

Fig. 5.2 Schematic representation of Y-shaped structure of an antibody

specificity of the diagnostic test. In conclusion, while polyclonal antibodies may have limitations, their high sensitivity and ability to detect a wide range of antigens make them an essential tool in immunology research and diagnostic testing [48, 50–52].

Immunosensors use immunoglobulins, also known as antibodies, as the biochemical receptor [32]. Antibodies, defensive proteins produced by an organism's immune system upon exposure to foreign biological substances, also known as antigens, create potent complexes with them. The detection of these antibodies and antigens involved in immunochemical interactions is carried out using Immunosensors. The existence of antibodies in the bloodstream may suggest either infection or the poisonous impacts of particular substances. Antigens are capable of being detected not solely in biological fluids but also in alternative surroundings, like the natural environment. Immunosensors can pick up on almost any substance given the presence of specific antibodies, demonstrating exceptional specificity and selectivity [53, 54].

5.4 Enzyme Bio-receptors

Enzymes are catalysts that are commonly known for speeding up biological reactions. They are natural proteins that are used in biosensors and catalyze specific molecules without being consumed in the process [55]. Enzymes possess the capability to identify a particular objective akin to a lock and key mechanism. These biosensors rely on the binding capabilities of enzymes and the catalytic reaction to identify the target analyte. An enzymatic biosensor can be created using either pure biological preparations with specific biological activity or enzyme preparations, such as tissue or microbial culture homogenates. To create the simplest design for an enzymatic biosensor, the product or substrate of the enzymatic reaction is made electrochemically active and capable of rapid and reversible oxidation or reduction using an appropriate potential [56].

Enzymes have been widely employed in biosensor applications as the most prevalent bioreceptor molecules due to their exceptional binding properties and impressive catalytic activity. Enzymes are often utilized as bio-receptors in biocatalytic detection mechanisms, which use catalytic reactions to enhance detection. All enzymes, except for a limited set of ribonucleic acid molecules that exhibit catalytic properties, are composed of proteins. Some enzymes require additional components, such as inorganic ions like Zn^2, Mn^2, Mg^2, Fe^2, or more complex organic or organometallic molecules known as coenzymes, to function properly. However, others only require their amino acid residues to function. Enzymatic activity has fewer limitations compared to other common methods [57–59]. Enzymes require their natural protein conformation to maintain their ability to catalyze reactions. If an enzyme is denatured, broken down into constituent amino acids, or divided into subunits, it loses its catalytic activity. Enzyme cascades frequently boost enzymatic activity, resulting in complicated cell processes [60]. Enzymatic receptors have the capacity to alter recognition mechanisms. This can be exemplified by the regulation of enzyme activity upon ligand binding to a receptor. Enzymes display heightened selectivity and sensitivity when contrasted with chemical processes, and they demonstrate relatively rapid response times in comparison to alternative biological sensors. Bioreceptors can use different mechanisms to perform their functions. Some Bioreceptors convert the analyte into a product that can be detected by a sensor, while others identify an analyte that acts as an activator or inhibitor of an enzyme [46, 61]. Enzymatic sensors can be categorized into two groups depending on their functions. The two categories are inhibitor sensors and substrate sensors. Substrate sensors are specifically designed to detect and identify particular substrates that are involved in the enzyme processes. For instance, glucose measurement can be easily conducted using a glucose oxidase-based enzymatic sensor, and the detection of urea can be carried out using a urease sensor. In contrast, inhibitor sensors are engineered to identify chemicals that decrease the activity of an enzyme, for instance, the identification of organophosphorus herbicides that obstruct acetylcholine hydrolysis mediated by acetylcholinesterase. The glucose and urea biosensors are the most commonly used among the various enzymatic biosensors [62–66].

5.5 Nucleic Acid Bio-receptors

The hybridization of DNA or RNA is another biorecognition process. Nucleic acid sequencing has been used for specific diagnostic applications since the early 1950s and is still expanding today [10]. Nucleic acid-based biosensors utilize oligonucleotides as sensing elements, which are composed of a specific sequence of RNA or DNA bases. These biosensors function through the precise hybridization of complementary DNA/RNA strands or by functioning as specialized receptors for biochemical/chemical species [67]. Nucleic acid-based biosensors utilize the specific and highly-affine binding process between two single-stranded DNA (ssDNA) chains to create double-stranded DNA (dsDNA) as the biological recognition element. This process is preferred due to its remarkable binding specificity, which allows for accurate detection and measurement of target molecules [5, 45]. Nucleic acid biosensors often use DNA probes or DNA primers as their receptors. The fragments used in DNA probes are not naturally occurring components isolated from living organisms, but rather are artificially produced to emulate the specificity of the complete DNA structure. The polymerase chain reaction is employed to produce DNA probes which can be altered to amplify their steadiness or expedite their integration into a biosensor. Aptamers are synthetic nucleic receptors that have no natural analog and are selected based on their ability to interact with certain biomolecules. However, since the required aptamer structure cannot be predicted, all possible oligonucleotides are synthesized (within reasonable limits of 40–100 nucleotides) and then the ones that bind most firmly to the target are selected [68–70].

The implementation of this technology has led to the advancement of DNA-based sensors and has surpassed conventional methods such as radioisotope coupling and electrophoretic separation, which are costly, time-consuming, and risky [71]. DNA-based biosensors are also known as Geno-sensors [45, 68]. Nucleic acid detection methods are the foundation for the development of quick and affordable infectious and genetic disease diagnostics. The development of DNA biosensors has garnered significant attention owing to their potential applications in forensic research, gene analysis, clinical diagnostics, and other medical arenas. Electronic biosensors called Geno-sensors can identify the individual nucleotides of DNA molecules that make up a genome. Automated Geno-sensors enable rapid and non-destructive DNA molecule sequencing. In contrast to antibodies and enzymes, nucleic acid recognition layers have the advantage of being rapidly generated and restored for various uses [72–75].

5.6 Cell-Based Biosensors

Antibodies, enzymes, and DNA are commonly used as bio-receptors in biosensing because they specifically target molecules. However, living cells present an exciting alternative to molecular bio-receptors due to their wide range of biomolecular mechanisms. This diversity of mechanisms can be harnessed for biosensing applications, making living cells a promising avenue for future research in biosensing technology [76, 77]. Cells exhibit a wide array of receptors in varying degrees, affording them the capacity to generate a quantitative reaction to distinct stimuli and the possibility of analyzing numerous analytes quantitatively with less expenditure and exertion (Fig. 5.3). In the past, the fundamental motive for using cells as bio-receptors was to achieve this goal. Utilizing cells as bio-receptors allows the necessary molecules and essential enzymes for biosensing to exist within their natural milieu, thereby maximizing the specificity and activity against the intended analyte. As a result, cells present a viable approach to designing the functional framework of biosensors, a feat previously unattainable by molecule-based biosensors. Meanwhile, larger eukaryotic cells have played a vital role in exploring fundamental cellular processes and the development of diseases. Ultimately, the power of these biosensors lies in their ability to link live cells to external transducers, creating a readable and quantifiable signal [76–78].

5.7 Electrochemical Biosensors

Electrochemical-based biosensors were the first biosensors to be scientifically designed and effectively marketed for various analytes. Several devices based on the electrochemical concept, including those for infections and poisons, have been suggested and marketed and are currently available [10, 79]. It is believed that electrochemical biosensors are poised to advance genetic testing in the future. An analytical detection process can demonstrate the electrochemical-based sensor mechanism through an enzymatic reaction, antibody-antigen interaction, whole-cell detection,

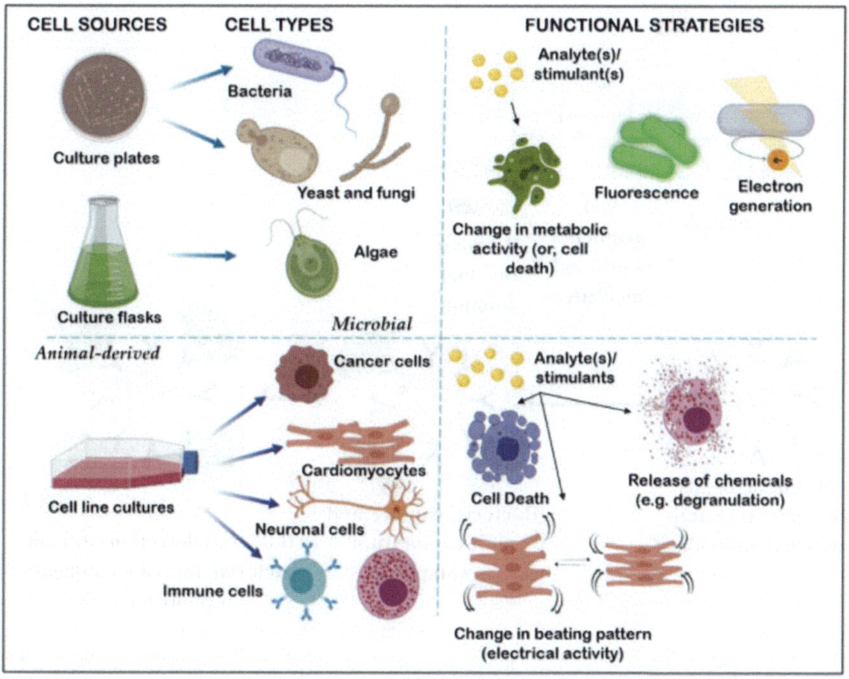

Fig. 5.3 Schematic illustrating the various cell types (on the left) and functional approaches (on the right) employed in cell-based biosensors [76]

or other means [80, 81]. In the realm of electrochemical biosensors, there exists a dynamic involving chemical processes that transpire between the targeted analyte and an immobilized biomolecule. These processes result in the creation or consumption of electrons or ions, thereby impacting the observable electrical features of the solution, such as potential or current. By leveraging the electrochemical signal produced, the analyte concentration in a given sample can be accurately assessed in a quantitative manner, An example of it is shown in Fig. 5.4 [17].

Electrochemical detection techniques are highly consistent with the development of biosensors due to their low cost, relatively simple instrumentation, high sensitivity, and low power requirements [2]. An excellent example of an electrochemically based biosensor is the disposable blood glucose strip (single-use) used for diabetic patient care. The enzymatic interaction with blood glucose in this biosensor produces H_2O_2, which is then electrochemically oxidized to provide a current for quantifying blood glucose levels.

There are three distinct kinds of electrochemical sensors that can be employed to identify variations in electrochemical reactions that arise throughout a process. These classifications are determined by the measured characteristic and involve impedance (impedimetric), potentiometric (potential), and amperometric (current) [2].

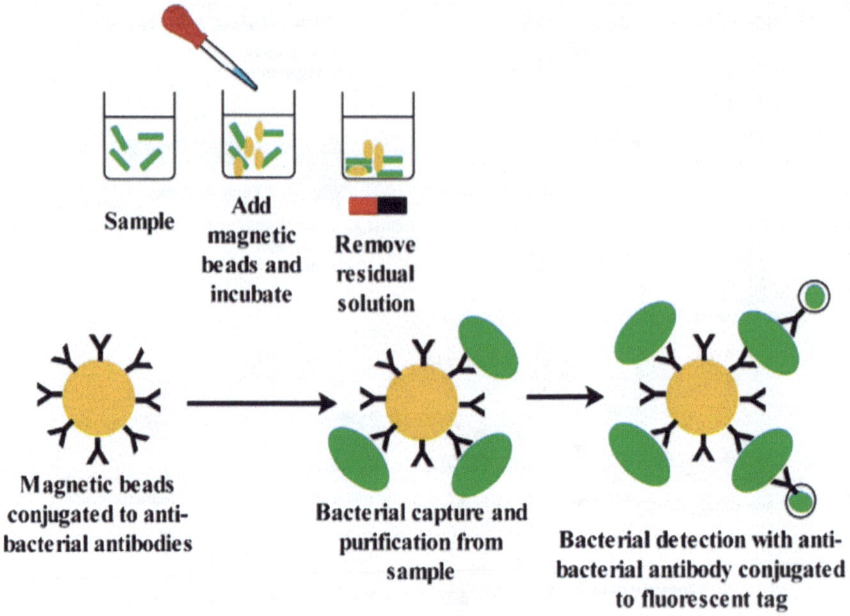

Fig. 5.4 A diagram illustrating the droplet-based electrochemical biosensor is presented, which is designed for the identification of E. coli. The biosensor employs magnetic beads that are bound to antibodies, serving as the biological recognition component [82]. Reprinted from sensors, MDPI journals

5.8 Optical Biosensors

Optical biosensors that detect various spectra, including refraction, SERS, Raman, phosphorescence, fluorescence, absorption, and dispersion spectrometry, are a diverse and multifaceted class of biosensors. Opto-sensors can utilize both label-based aptamers (incorporating nanoparticles, enzymes, luminophores, and fluorophores) and label-free detection systems such as surface plasmon resonance. In optical bioassays, aptamers have been extensively employed as biorecognition elements, with colorimetry and fluorescence being the two most critical techniques [32, 37]. These methods of analyzing spectra can measure various properties, such as phase, decay time, amplitude, polarization, and energy. Amplitude is the most frequently measured property since it is easily correlated to the concentration of the analyte of interest. Optical biosensors, which employ optical fibers, can detect analytes based on light scattering, fluorescence, or absorption. The beauty of optical biosensors lies in their non-electrical nature, which renders them suitable for in vivo applications, and their ability to detect multiple analytes using different monitoring wavelengths. The adaptability of fiber optic probes stems from their capacity to transmit signals that reveal alterations in the distribution of the spectrum wave propagation, intensity, time, wavelength, or polarity of the light [83]. Optical biosensors consist of a light source and various optical segments that work together to produce a light beam with particular properties, directing that light to a balancing operator, an adjusted detecting head, and a photodetector. Signals are typically received from these devices using flexible cables that allow light to be transmitted to the biological component. Optical techniques are highly versatile and enable multiple wavelengths to be examined simultaneously without any interference. Techniques including interferometry, fluorescence spectroscopy, surface plasmon resonance, and spectroscopy of guided modes in optical waveguide structures such as grating pairs and resonant mirrors—are often used in biosensors. On the other hand, other emerging optical sensor methods, such as photonic crystals and optical ring resonators, are of interest [84, 85].

5.9 Piezoelectric Biosensors

Piezoelectric materials are widely utilized in various technological fields and are commonly incorporated into electronic devices. These materials operate as oscillators based on the piezoelectric effect principle. They are easily detectable on their surface interactions, making them ideal for constructing biosensors that recognize affinity interactions. Such studies are well-suited for assays that rely on the interactions between an antibody and an antigen, aptamer, protein, and two polynucleotide strains [86]. The piezoelectric biosensor is a fascinating type of microelectromechanical system (MEMS) that operates by detecting changes in the resonance frequency of an oscillating crystal due to the interaction between an analyte and a bioreceptor. In this type of biosensor, the transducer is made of piezoelectric material, typically quartz, and a biosensing material is applied to it, causing it to vibrate at its natural frequency. The piezoelectric effect is a physical phenomenon in which a material produces a voltage when subjected to mechanical stress, and it also works in the opposite direction [87–89].

When a piezoelectric material's surface is subjected to an alternating voltage, it generates mechanical stress or oscillation. Piezoelectricity is commonly observed in anisotropic crystals, which lack a center of symmetry. Piezoelectric materials play a crucial role in the construction of physical sensors and biosensors, making this method highly attractive in analytical chemistry. Various materials such as polyvinylidene fluoride, tartrate tetrahydrate (Rochelle salt), quartz (SiO_2), gallium orthophosphate, aluminum phosphate (berlinite), lead titanate, barium, crystalized topaz $Al_2SiO_4(F, OH)_2$, zinc oxide, aluminum nitride, and polylactic acids are commonly used for this purpose. The frequency of the crystal, when coated with biosensing material, is dependent on both the mass of the coating and the crystal. By measuring the resonant frequency with precision, it becomes possible to determine the mass of the analyte adsorbed on the crystal surface [88, 90]. To activate the biosensor or sensor, one must apply an alternating voltage to the surface by two electrodes. This voltage will result in mechanical oscillations in the crystal. Afterward, the oscillation frequency is determined by inserting the crystal into a circuit that oscillates. The presence of any mass, including analytes, on the surface of the electrodes will lead to a change in oscillation frequency. According to Sauerbreay and colleagues, the discernible variation in frequency, commonly known as decay, is an accurate indication of the weight of mass attached to the crystal. However, it is important to note that variations in medium viscosity can also impact the oscillations, as explained by Kanazawa and coworkers [91]. Piezoelectric sensors, which are categorized into two main types, namely surface acoustic wave (SAW) and bulk wave (BW), have been deemed less popular and inferior in comparison to electrochemical and optical-based biosensing techniques. Nonetheless, there are certain papers in the publications that reveal the utilization of piezoelectric sensors in various applications, including the detection of hepatitis C and B, cholera toxin diagnostics, and even the recognition of foodborne pathogens [92–95]. Piezoelectric immuno-sensors have been recognized as one of the most sensitive types of sensors created thus far. They are proficient in detecting antigens within the picogram range, as reported in the reference [89, 96].

These sensors work by coating the piezoelectric material with a bioreceptor, such as an antibody or aptamer, that specifically recognizes the target analyte. The binding of the objective analyte to the bioreceptor leads to a modification in the oscillation frequency of the piezoelectric sensor, which can be measured to quantify the target analyte. In summary, while piezoelectric sensors have historically been considered less popular compared to other technologies, they are still capable of detecting a wide range of substances with high sensitivity. Piezoelectric immuno-sensors, in particular, have proven to be highly accurate and precise in detecting low levels of antigens.

5.10 Thermal Biosensors

Sensitive thermistors can be utilized to measure heat and ascertain the concentration of an analyte, rendering them valuable tools in biosensing. Although thermal biosensors are not widely used, a wide variety of biochemical components can be integrated with different transducers to create a diverse range of biosensors. The majority of biosensors, around 80%, are electrochemical in nature and can be classified as enzyme electrodes, immuno-sensors, or DNA sensors depending on the biological component utilized [46, 97, 98].

Enzymes are highly specific and many of their reactions release heat, making them a useful component in thermal or calorimetry-based biosensors. Thermal biosensors are able to detect reactions by observing alterations in enthalpy levels. After that, calibration curves are created using the data collected. The methodology of thermal measurement presents an exceptional opportunity for the identification of various biomolecules, and it complements the functionality of other biosensor detection systems in a remarkable manner. Thermal biosensors are a remarkable example of cutting-edge technology that employs a flow injection analysis system to integrate an immobilized enzyme reactor and a differential temperature measurement configuration. These biosensors utilize thermal transducers, such as thermopiles or thermistors, that are strategically placed along the enzyme Foundation. This measurement setup effectively minimizes temperature fluctuations, thus allowing for precise measurement of enzyme catalysis. The thermal signal, which is directly linked to the substrate concentration, offers a considerable benefit when compared to indirect methods that rely on electrical or optical detection. With a growing demand for biosensors that offer heightened sensitivity, enhanced reproducibility, and broader dynamic ranges, the continuous advancement of biosensor technology remains an essential area of focus. Thermal biosensors have been widely adopted in various fields, including clinical analysis, environmental monitoring, bioprocess monitoring, biochemical studies, and organic solvent reactions [99].

5.11 Biosensors-On-Chip

The merge of biosensors with microfluidics systems in lab-on-chip (LOC) technology has become increasingly popular due to the numerous benefits it offers. This approach provides a miniaturized and integrated alternative to traditional laboratory methods, which can be repetitive and time-consuming. Microfluidics pertains to handling and controlling small amounts of fluids within microchannels, and the integration of various detection tools has made it an attractive platform for biosensors. Together, microfluidics and biosensors have enabled the development of miniaturized devices that can process samples and detect analytes with high sensitivity and quantification. This combination is particularly useful in the realm of point-of-care diagnostics, where robust and cost-effective biosensors-on-chip are needed [100]. We believe that the integration of microfluidics and biosensors is essential for the advancement of point-of-care diagnostics, and has the potential to revolutionize healthcare by offering faster, more accurate, and more accessible diagnostic tools.

5.12 Continuous Microfluidic-Based Biosensors

Continuous microfluidic biosensors are built of narrow channels that manipulate the continuous flow of fluid utilizing external sources such as integrated micropumps, mechanical pumps, pressure sources, or internal processes such as a mix of electrokinetic forces and capillary. Continuous microfluidic chips can rigidly control the volume up to nanoliters (NL), including different molecules (cells, proteins, DNA, and so on). The implementation of this technological advancement has a diverse range of uses in fields such as biochemical and chemical separation, as well as in micro total analysis systems (μ-TAS) and micro-reactors. It enables the execution of a comprehensive analytical process, starting from sample pre-processing, all the way through to detection [101, 102]. When there is a need for a higher level of adaptability or more intricate fluid handling, the performance of the application is impacted, leading to multiple difficulties. For instance, it isn't easy to integrate closed channel systems efficiently, so that system characteristics do not affect fluid flow in different regions. Furthermore, the microstructures undergo a permanent etching process, which puts a restriction on their ability to be reconfigured and achieve fault tolerance. Since 2000, continuous microfluidic devices have advanced dramatically, allowing for significant research of biological systems, especially at the molecular level. Features of these systems include multisensory platforms and sample size reduction, energy consumption reduction, reagents cost reduction, waste generation reduction, and overall measurement times reduction. Microfluidics has lately been shown to be helpful in various applications, including biomedical applications such as point-of-care devices, clinical diagnostics, chemical and systems biology, environmental monitoring and evaluation, biological screening, drug discovery, and more [25].

5.13 Droplet Microfluidic-Based Biosensor

The application of microfluidics technology based on droplets offers great potential for biosensor development. As previously mentioned, droplet microfluidics is recognized for its aptitude to produce droplets with high throughput and utilize each droplet as a miniature chemical container. Such microfluidic systems produce and use a flow of sub-nanoliter droplets dispersed with a non-miscible continuous phase, which permits experimentation with ultra-high throughput and the capacity to simulate conditions that resemble those of a single cell. This feature allows for the categorization of chemical and biological reactions. Microfluidic systems are capable of performing digital fluidic activities using fluid plugs or droplets that contain biological and chemical reagents. These systems have enabled precise control and manipulation of small-scale fluidic processes with a high level of accuracy and efficiency [25, 103, 104].

Plug-based microfluidics is a method that entails suspending nanoliter droplets of fluid within an immiscible fluid to facilitate mixing, merging, splitting, dispensing, and transporting of aqueous droplets in microfluidic channels. These miniature reactors have been effectively used as small PCR compartments, reducing sample volume requirements and limiting diffusion. One of the advantages of these systems is their capacity to sustain droplets within the immiscible fluid for extended periods without evaporation. This feature permits the preloading of nanoliter plugs of reagents for the incubation of cells or high-throughput screening of precisely controlled droplets before analysis [105–108]. Droplet-centered microfluidic systems present an array of benefits for expedited screening and delicate bioassays, particularly PCR. By utilizing a variety of techniques like needles, electric fields, micro-injectors, and electric fields, Microdroplets are formed within microfluidic channels. When aqueous reagents are introduced into an immiscible carrier fluid like oil, a remarkable phenomenon occurs—the spontaneous emergence of droplets, leading to the formation of a water-in-oil (w/o) emulsion. This is achieved through the ingenious application of T-junctions or flow-focusing geometries. The design of droplet-based microfluidic systems provides enhanced mobility and reduced energy requirements. Furthermore, the dependable performance and effectiveness of these devices have rendered them advantageous for specialized applications such as defense, security, environmental evaluation, and clinical diagnostics [109, 110].

5.14 Digital Microfluidics (DMF)

DMF is an effective microfluidic tool that is utilized for the control of complicated bioassays. Through digital microfluidics (DMF), microdroplets can be managed; this system architecture demonstrates adaptability and can be easily reconfigured. Most importantly, it successfully safeguards against sample contamination throughout the assay. Electrowetting, commonly referred to as EWOD, enables the creation of unique droplets that can be controlled independently and with ease. This innovative technique, which bears some resemblance to digital microelectronics, is known as "digital" microfluidics. In integrated devices, digital microfluidics enables the manipulation of liquid samples in discrete droplets using an array of electrodes coated with a dielectric material, as depicted in Fig. 5.5. Through the application of a sequence of potentials to the electrode array, the droplets can be individually or simultaneously mixed, merged, split, and dispensed, all programmatically controlled. Digital microfluidics (DMF) technology can be implemented using either a two-plate or one-plate device format. It boasts the capability to handle droplets of varying sizes, ranging from Picoliters to Microliters, leading to reduced consumption of reagents, akin to other microfluidic systems [111–113].

The digital microfluidic platform offers the added benefit of flexibility, also referred to as dynamic reconfigurability. This is because fluidic processes can be executed at any point on the array. The programming of droplet routes and operation scheduling is carried out by a microcontroller that propels electrodes in the array.

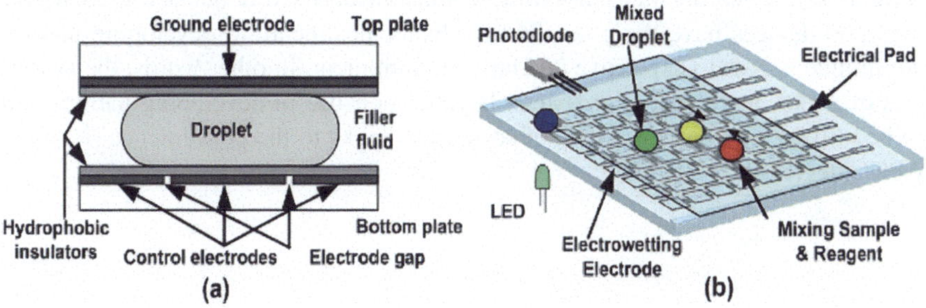

Fig. 5.5 **a** The fundamental cell employed in a digital microfluidic device; **b** A visual representation of the microfluidic array from the top perspective. Reprinted from sensors, MDPI journals [111]

Moreover, digital microfluidic arrays incorporate optical detectors like photodiodes and LEDs to oversee colorimetric bioassays [114]. Valves and pumps are not necessary for flow control in these systems, as they leverage platforms that are compatible with the widely adopted biosensor detection methods, such as electrochemical and optical techniques. The integration of digital microfluidics (DMF) has played a significant role in advancing the development of numerous biosensor devices [115].

5.15 New Developments in Microfluidic Platforms

The microfluidic platforms comprise crucial elements, including the constituent materials and the techniques employed to regulate fluid flow. Although microfluidics offers immense potential in POC medical diagnostics, certain constraints impede its applicability. First and foremost, manufacturing costs and material compatibility are two essential factors to consider while developing new materials. Additionally, fluid control methods necessitate the use of intricate and costly off-chip components that are not easily employed beyond the laboratory premises. In recent years, new materials have been developed to overcome some of the limitations of microfluidic chips. These materials include engineered plastics such as photocurable perfluoropolyether (PFPE) and cyclo-olefin copolymers (COC), as well as paper. Unlike PDMS chip fabrication, the production of thermoplastic chips or paper-based chips does not require the use of a clean-room infrastructure for master construction. This characteristic makes them exceptionally well-suited for large-scale implementation, providing a substantial advantage in terms of commercialization. Furthermore, these innovative technologies offer numerous benefits for point-of-care (POC) diagnostics. The first advantage of paper-on-chip devices is their cost-effectiveness compared to silicon and glass, resulting in inexpensive materials and fabrication methods. This enables large-volume manufacturing at a low cost. Secondly, the disposability of these chips eliminates the risk of cross-contamination of biological samples, which is a significant benefit due to their affordability. The single-use nature of these devices eliminates the need for maintenance or recalibration. Thirdly, both plastics and paper are lightweight, making them easy to store and transport. Consequently, these newly developed devices have the potential to enhance healthcare in developing nations with limited access to expensive and bulkier equipment. In other words, the advantages of paper-on-chip devices extend beyond their use in developing nations and can be utilized in a broader range of scenarios thanks to the convenience provided by POC devices [25, 26, 116].

5.16 Paper-Based Chips

The field of microfluidics has undergone a transformative change with the introduction of paper-based technology, which offers a cost-effective alternative to traditional materials like PDMS. This advancement has unlocked numerous previously unattainable applications. A notable advantage of paper-based diagnostic tools is their ability to function without the need for external power sources, equipment, or sample processing. These systems rely on capillary actions and fluid wicking into the inter-fiber space of the paper to facilitate fluid movement. However, commercially available paper strips for diagnostics, such as immunochromatographic strips, have limitations. They cannot perform quantitative assays and lack the ability to conduct multiplexed analyses. Nevertheless, incorporating multiplexing can enhance the screening process, which is a common practice in the field. The emergence of paper microfluidics has gained wide acceptance due to its disposability, flexibility, ease of fabrication, low cost, and capillary-driven liquid transport without the need for external driving mechanisms. These benefits have inspired scientists to develop point-of-care instruments for detecting illnesses in resource-limited regions [117, 118]. However, the use of paper-based chips is currently limited to proof-of-concept studies and rapid prototyping due to certain drawbacks that hinder broader application. These limitations include incomplete delivery of the original sample volume to the testing area, with only around 50% reaching the desired location. Additionally, the narrow working channel widths, typically $\geq 500\ \mu m$, pose challenges as they can become obstructed when exposed to high temperatures, such as the melting point. The amalgamation of biosensors and microfluidic systems yields a formidable analytical instrument that presents a remarkable stride toward domestic diagnostic techniques, mutually advantageous for both burgeoning and established nations. This integration provides numerous favorable advantages, including disposability, portability, high detection accuracy, minimal handling of hazardous materials, cost-effective analysis, reduced processing time, automated sample preparation, a significant reduction in waste production, and low reagent consumption [119, 120].

Microfluidic biosensors offer several advantages over regular detection methods. They not only reduce costs but also improve detection sensitivity limits and specificity. Despite their small size, a single microfluidic biosensor can perform a comprehensive analysis, which includes continuous sampling, pre-concentration, sample separation, treatment, and blending. Additionally, these biosensors provide enhanced analytical capabilities, immediate detection, increased throughput, rapid response times, and mobility. These features make them highly adaptable for point-of-care (POC) applications [121]. In this particular scenario, point-of-care (POC) instruments hold immense potential for rapid diagnosis, prevention, and therapy of various human ailments. The need for faster development of point-of-care devices to facilitate effective therapeutic intervention has become urgent due to the emergence of new or recurring infectious diseases, as well as lifestyle changes. Additionally, routine biochemical tests require highly skilled technicians and costly laboratory infrastructure, and a consistent supply of chemicals, all contributing to escalated expenses.

In order to tackle the issues arising from the diagnosis process, novel technologies have been created to ensure that testing equipment is easily accessible to patients. These pioneering breakthroughs, also known as point-of-care (POC) diagnostics, comprise analytical testing systems that can be utilized both in the clinical laboratory and on the go through the utilization of portable equipment. As a result, there has been a notable transformation in the clinical diagnosis and treatment field. The shift from conventional diagnostic instruments to compact and user-friendly point-of-care (POC) devices has been remarkable. These cutting-edge gadgets fulfill the growing need in the medical field for automated tools capable of directly using unprocessed materials. Additionally, they need minimum upkeep, whether mechanical or electrical. Point-of-care (POC) diagnostics can provide fast and precise outcomes at cost-effective rates, particularly for patients requiring urgent biochemical testing. An ideal diagnostic device should provide an optimum setting in which the material being examined in a sample may engage with the bio-recognition molecule (such as cells, DNA, lipids, or proteins) that is fixed in place on the sensor surface. Previously, the creation of such devices seemed to be unattainable because of the substantial inconsistency in biological samples, resulting in incorrect clinical data. Nevertheless, current endeavors are concentrated on enhancing the exactness and accuracy of analyte detection techniques inside the integrated unit for sample preparation. In the context of medical diagnostics, microfluidics devices have the potential to yield significant advancements. The latest progress made in the fields of microfluidics, nanomaterials, sensors, and arrays has effectively facilitated the development of "lab-on-a-chip" instruments. The world is currently encountering fresh healthcare challenges that have stimulated novel technological advancements by merging diverse fields. To address these present-day obstacles, there is a need for timely and precise identification of various lethal illnesses through the creation of new biosensors, as well as the employment of integrated systems such as microfluidic chips that can handle multifaceted samples efficiently and cost-effectively with minimal training. The merging of microfluidics and biosensors, two well-established and distinct domains, has resulted in viable solutions for critical healthcare concerns [121–123].

5.17 Microfluidics-Based Biosensors for POC Diagnostics

Biosensors have become a ubiquitous device in point-of-care (POC) diagnostics. By integrating them with microfluidics, there has been a proliferation of diagnostic applications.

Currently, there is a significant emphasis on advancing a wide range of microfluidics-based biosensors that hold the potential to enhance diagnostic efficiency. These biosensors include non-invasive biosensors, microarrays, optofluidics, immuno-sensors, and enzymatic biosensors. In the subsequent sections, we will provide a comprehensive overview of the research and development that has been conducted on these biosensors in recent years [122].

5.18 Glucose Biosensor

The typical course of action for treating diabetes involves precise and diligent observation of glucose levels within the body. This is accomplished through the use of a portable glucose meter that tracks blood levels. The device itself is composed of an enzyme-activated test strip that is disposed of after a single use. In order to activate the strip, a small drop of blood from a puncture in the finger is applied to it [124]. Since the first glucose biosensors were introduced in 1962, several technologies have been developed for the electrochemical bioassay of glucose. Most of these devices are built on disposable electrode strips printed with enzymatic functional screens that use the principle of amperometric detection. The creation of manual POC testing equipment can help monitor itself and obtain fast, reliable, and accurate findings from small sample volumes. The examination of glucose levels through a biosensor typically employs two mechanisms, as depicted in Fig. 5.6 [124].

Several techniques have been employed to improve the features of glucose-sensing systems through their integration with microfluidics. Hou et al. have presented a method whereby a glucose microfluidic chip is fabricated through CO_2 laser and hot plate press bonding techniques on polymethyl methacrylate. This approach is noteworthy for its innovative and practical aspects. This incredible chip presents an unparalleled benefit compared to standard detection techniques as it boasts an additional advantage of rotation. The spinning action generates an increased level of mixing within the micromixer and can be harnessed for the swift and cost-effective detection of glucose. Recent breakthroughs in the fields of engineering and nanotechnology have paved the way for the immobilization of enzymes onto nanostructures and nanoparticles, thereby enabling their functionalization onto the designated electrodes. As an example, Yu et al. successfully created a cutting-edge microfluidic chip that involved the immobilization of enzymes onto the single-walled carbon nanotubes

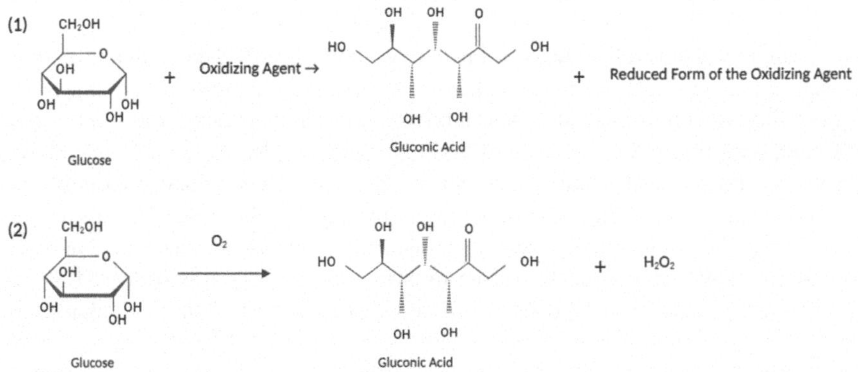

Fig. 5.6 Schematic representation of biochemical reactions used in enzymatic biosensors (glucose biosensors). (1) Catalytic conversion of glucose to gluconic acid in the presence of a catalyst on the electrode surface. (2) catalytic conversion of glucose to gluconic acid and H_2O_2, after which electrons are produced through electrochemical oxidation

of the PDMS channel. This microfluidics device, crafted through photolithography, has a robust linear response of 5 × 10–3 mol/L. Moreover, the photolithography technique has also been utilized to manufacture a paper-based microfluidic glucose sensor utilizing Whatman filter paper. Nevertheless, there remains a pressing need to enhance the shelf life and linearity of these paper-based glucose-sensing devices before commercialization can occur [24]. While this methodology yields accurate results, the need for frequent testing throughout the day can lead to patient dissatisfaction and discouragement. The continuous perforation of the skin for testing purposes has resulted in persistent pain, the development of scars, and decreased sensitivity in the fingers. The emergence of tears as a fascinating alternative for glucose monitoring has captured the attention of medical practitioners. This bodily fluid exhibits a promising relationship with blood glucose levels, presenting an opportunity for less invasive analysis in patients. However, selecting a suitable method for determining glucose levels in tears necessitates careful consideration of factors such as selectivity, a low detection limit ranging from μmol L^{-1} to mmol L^{-1}, and a minimal sample volume of just a few μL. Due to the use of diminutive detection platforms, especially microfluidic devices, and contact lenses, these investigative techniques have made notable strides recently. Although encouraging, numerous proposed contrivances are costly and entail complicated production procedures that require highly specialized equipment. The sole inexpensive devices are disposable components, although the overall cost rises when many analyses are necessary [125, 126].

Utilizing microfluidic devices based on cotton threads may efficiently overcome the restrictions provided by glucose analyzers. These platforms possess unique attributes that enable the creation of simple and affordable devices. In comparison to conventional materials used in miniaturized platforms, such as polymers, glass, and silicon, cotton threads are more accessible and cost-effective. Moreover, cotton threads exhibit a remarkably hydrophilic nature due to their high surface energy. The abundance of spaces between the fibers enables the transportation of solutions through capillary force, eliminating the need for any mechanical pumping mechanism. This investigation has developed a rapid and cost-effective technique for measuring glucose levels in tears. It involves combining a PTB-GOx amperometric biosensor, a μTED (micro total analysis system), and measurements using μFIA (micro flow injection analysis). The fabrication of the biosensor was simple and efficient, requiring only one stage of 5-voltammetric cycles. One notable achievement was the successful immobilization of PTB and GOx (glucose oxidase) on the biosensor surface. This immobilization ensured that there was no loss of the film during the flow analysis, and it maintained a consistent enzymatic activity over time. However, a gradual decline in enzymatic activity was observed as the analysis progressed. The PTB-GOx biosensor exhibited an impressive characteristic of long-lasting endurance, maintaining a remarkable 70% response rate even after a thirty-one-day analysis. The μTED, constructed using inexpensive and readily available materials, featured a straightforward fabrication process and demonstrated exceptional reproducibility in its responses across different devices (RSD = 3.13%). Thanks to the groundbreaking μFIA technique, a fresh sample could be rapidly injected every thirty seconds without the need for any component replacement. This was made possible by the continuous transport of solutions within the system, the swift purification of the biosensor surface, and the outstanding stability between injections (RSD = 2.54%) [126–128].

The methodology presented in this study aims to simplify the analysis of glucose in tear samples from individuals without diabetes. It eliminates the need for purification and requires only a small sample volume (a few microliters). The obtained results have shown a strong correlation with blood glucose levels, highlighting the potential of this method as a non-invasive and cost-effective approach for indirectly monitoring blood glucose through tear analysis. Furthermore, there is potential for further improvement in this tear glucose analysis method. By incorporating a compact battery-powered electronic device and a smartphone application into the analysis system, it could achieve full portability. This advancement would enhance its usability, enabling convenient monitoring of blood glucose levels anytime and anywhere. Additionally, the immobilization of other enzymes in the PTB film will be explored to enable a selective analysis of different biomarkers present in tears.

5.19 Cholesterol/Lipid Biosensor

Mammalian cell membranes, vitamin D, cell signaling, and hormones all depend on cholesterol as a vital structural component. Cholesterol is also essential for maintaining proper membrane fluidity and permeability. In plasma, cholesterol is transported by lipoproteins, including high-density lipoprotein (HDL), intermediate-density lipoprotein (IDL), low-density lipoprotein (LDL), and very low-density lipoprotein (VLDL). Among these, LDL, which is cholesterol-rich, is commonly referred to as 'bad' because it contributes to the accumulation of cholesterol in the arterial walls, leading to various heart and vascular diseases [129]. Therefore, it is necessary to measure cholesterol levels to evaluate a patient's overall health. The Friedewald equation is used to estimate the lipid content indirectly, as LDL levels cannot be measured directly [122]. To estimate free cholesterol, Ali et al. developed a microfluidic biosensor that utilizes nanostructured anatase titanium dioxide. In their study, Wisitsoraat and colleagues introduced a microflow injection analysis (μFIA) system that incorporates functionalized carbon nanotubes, enabling real-time monitoring of cholesterol [130]. Ruecha and colleagues developed a PDMS-based microsystem that utilizes electrochemical techniques for the rapid detection of cholesterol [131]. Ali and his fellow researchers have developed a microfluidic nano biochip that demonstrates a remarkable level of efficiency. This biochip utilizes nanostructured nickel oxide to effectively monitor total cholesterol levels with high effectiveness. However, it can be challenging to stabilize cholesterol oxidase and cholesterol esterase on an electrochemical transducer platform for an extended period, despite the high sensitivity and precise detection of cholesterol offered by macro/microelectrodes [130]. Efforts should be made to address the issues concerning the movement and control of fluids in cholesterol microfluidic chips equipped with syringe head pumps, and suitable solutions should be sought. Furthermore, it is necessary to develop microfluidic-integrated biosensors for the detection of LDL. Paper-based cholesterol biosensors offer a promising solution for diagnostic purposes, particularly in resource-limited regions, aligning with the guidelines established by the World Health Organization (WHO). These biosensors need to meet the ASSURED criteria, which include being equipment-free, robust, rapid, user-friendly, specific, sensitive, deliverable to end-users, and affordable. Paper-based biosensors possess desirable properties such as liquid transfer without external equipment, capillary action, disposability, lightness, and thinness. These characteristics, combined with the ability to modify paper with diverse chemicals and create microchannels, make the paper an excellent material for the development of innovative biosensing devices. Because of its sensitivity and low detection limits, electrochemical detection is a promising choice for cholesterol biosensors. Because of this property, it is ideal for use in developing handheld diagnostic devices for use at the point of care (POC). Electrochemical biosensors may be easily integrated with flexible substrates, such as paper. The paper or a separate layer connected to the microchannels may be used to form the electrodes. Compared to other types of analytical biosensors, such as those that utilize optical technology or acoustic, electrochemical sensors require less instrumentation, making them more practical for field use. By utilizing paper patterning techniques like laser treatment, inkjet etching, wax printing, or photolithography, hydrophobic and hydrophilic areas can be created, forming channels bounded by hydrophobic barriers. A highly successful development of a paper-based microfluidic biosensor for electrochemical sensing of cholesterol was achieved. The biosensor utilized a working electrode composed of NiO nanoparticle-modified graphite, and notably exhibited a high level of sensitivity, measuring 26 mA mM^{-1} cm^{-2} within a wide linear range of cholesterol concentrations, ranging from 0.12 mM to 10.23 mM. These promising findings highlight the potential of this biosensor for point-of-care diagnostics [132].

5.20 Urea Sensor

The ingestion of proteins by the human body initiates the process of hydrolysis within the digestive system, leading to the eventual breakdown of said proteins into individual amino acids. Within the primary metabolic process known as the urea cycle, the liver plays a crucial role. As a byproduct of this cycle, urea is directly released into the bloodstream by the kidneys and subsequently excreted in urine. The normal range for urea concentration in the body of a healthy individual typically falls between 2.5 mM and 7.5 mM. Any deviations from this range can potentially indicate the presence of kidney disease and hyperammonemia. Furthermore, increased levels of urea can contribute to a number of conditions such as shock, dehydration, gastrointestinal bleeding, renal dysfunction, and urinary tract obstruction [133, 134]. The determination of urea levels in biological samples frequently involves the utilization of urease and glutamate dehydrogenase enzymes. Urease operates by transforming urea into ammonia and hydrogen bicarbonate, while glutamate dehydrogenase transforms ammonium ions, nicotinamide adenine dinucleotide (NADH), and α-ketoglutarate into L-glutamate and NAD+, as illustrated in Fig. 5.7. To determine the concentration of urea, the current produced by the electrons evolved after converting NADH to NAD+ is assessed. The demand for microfluidics-based urea biosensors has increased due to high accuracy, portability, fast reaction time/response, high sensitivity, reusability, and low consumption of reagents/materials [134].

The covalent immobilization of enzymes onto a gold electrode through the utilization of a self-assembled monolayer consisting of 10-carboxy-1-decanthiol has been successfully accomplished. However, it has been observed that the resulting urea sensor, which is based on this particular method, exhibits limitations with regard to its sensitivity and the minimum concentration it can detect. To mitigate these issues, an alternative approach has been explored. Notably, through the immobilization of enzymes onto a titania-zirconia nanocomposite, remarkable enhancements in both sensitivity and the range of concentrations that can be detected by the urea sensor have been observed. Urease immobilization can also be accomplished through the use of nanomaterials. The distinct chemical reactant properties of titanium oxide (TiO_2) and nickel oxide (NiO) have led to their development as matrices in electronic urea biosensors. NiO offers advantages such as fast electron transfer capability, and chemical stability making it a viable option. Additionally, it has been utilized as a sensing material for the detection of urea, uric acid, and glucose. These studies collectively attest to the promising nature of NiO and TiO_2 when utilized as sensing matrices. It is noteworthy that the use of NiO and TiO_2 in urea sensors is uncommon, and as such, we have employed them as the sensing matrices in our urea biosensors. The scientific community has recently focused on investigating the simultaneous detection of urea and its application in microfluidics. This research has highlighted

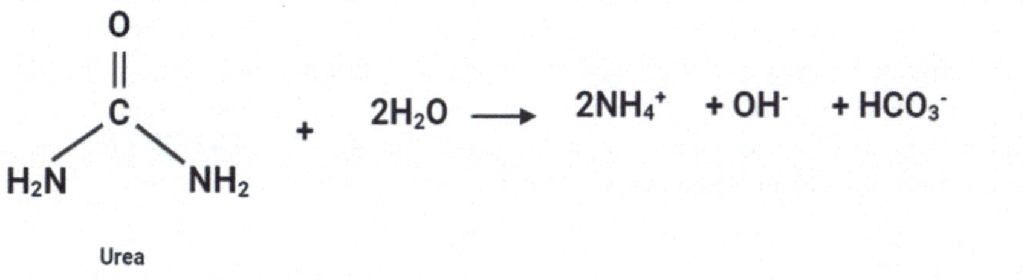

Fig. 5.7 Biochemical reaction in urea biosensor

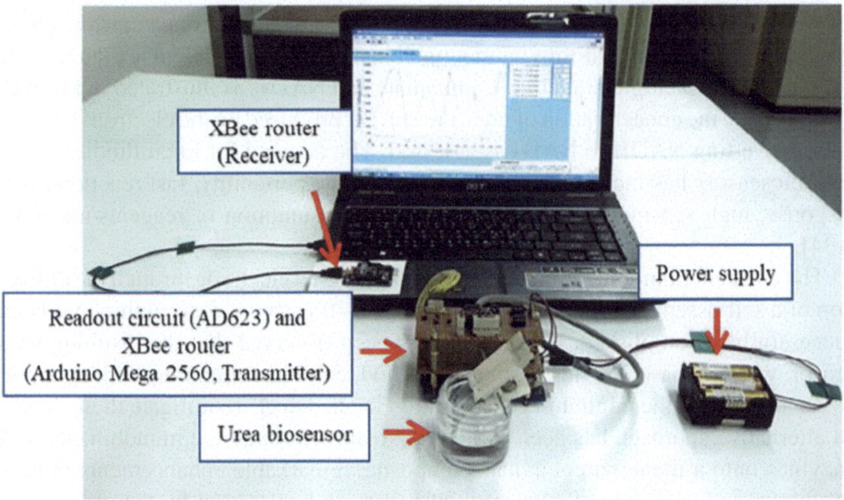

Fig. 5.8 Setting up a wireless measurement system for measuring urea. Reprinted from sensors, MDPI journals [135]

the potential of using TiO_2 or NiO as sensing matrices, as they facilitate electron conduction and improve enzyme binding. In this study, we aim to delve deeper into the sensing properties of urea biosensors integrated with wireless and microfluidic measurement systems, building upon previous findings in this area. Microfluidic systems are commonly used due to their fast response time, minimal reagent usage, and automation for continuous analysis of specific substance concentrations. To ensure continuous analysis and on-the-spot testing for urea, researchers conducted a study on how the urea biosensor responds to microfluidic flow conditions. They aimed to remotely detect urea levels by integrating the biosensors with a wireless sensing system using the ZigBee standard. The system successfully measured the concentration of urea and displayed the results in real time on a screen, enabling swift diagnostic procedures. Figure 5.8 illustrates the wireless measurement setup, which consists of a urea biosensor, a computer running the LabVIEW program, a readout circuit, and an XBee module [134].

5.21 Optofluidics

The integration of photonics and microfluidics at the micrometer scale has emerged as a promising field of research, known as optofluidics. By employing microfluidics for microscopic optical imaging and label-free non-contact measurements, scientists have developed a highly effective technique for real-time detection of biomolecules with exceptional sensitivity. Optofluidics focuses on manipulating light and fluids at the microscale in order to produce adaptable systems. The integration of optical devices into microsystems through 'lab-on-a-chip' and microelectromechanical systems (MEMS) communities has led to the development of a diverse range of optofluidic devices. Recent attempts to combine micro photonic devices and microfluidic systems have created new opportunities. These can be broadly classified into two categories. In the realm of integrated biochemical sensing, fluids can be transported through highly sensitive micro photonic circuits for analysis. Alternatively, microfluidics can be used to control micro photonic devices, adaptive, reconfigurable, and making them tunable. The ultimate goal of photonics is to miniaturize devices and integrate numerous multiple optical components onto a small-scale chip, thereby creating photonic integrated circuits that are both cost-effective and highly functional [136, 137]. Several effective biosensing platforms have been discovered for detecting and quantifying specific biomolecules. These include fluorescence tags, optical microscopic imaging, and surface plasmon resonance-based detection. Additionally, internal processes like cell division and migration may now be seen using microfluidic technologies. The progress of nanoparticle-based biosensors, whether in a lab or within an organism, has been greatly aided by the invention of several nanoparticulate imaging contrast agents. As an example, Measor et al. showcased the use of liquid-core anti-resonant reflecting optical waveguides for the on-chip detection of oligonucleotides. A detection limit between 10 nM and 10 μM was achieved by using a particular optofluidic filter design in conjunction with fluorescence-tagged molecules (Cy3 and Cy5) at a concentration of 600 nM. Combining a microscope with optofluidic channels holds the potential to unlock new possibilities for point-of-care applications [137].

5.22 Dango Virus Diagnosis

Microfluidic systems are miniaturized devices that integrate multiple functional modules onto a single chip for precise control of fluid flow. These systems have shown promise in enabling faster and more accurate diagnosis of viral infections such as dengue by manipulating and analyzing very small fluid volumes. Dengue viruses (DENVs) are a type of arthropod-borne virus (Arbovirus) that are transmitted by various Aedes mosquitoes. These mosquitoes can also carry other viruses such as encephalitis virus, yellow fever virus, and West Nile virus. The virus itself has a size of approximately 50 nm and a spherical shape. Its genetic composition is made up of a single positive sense RNA strand, spanning an impressive length of 11,000 nucleotides. It contains only one open reading frame [138, 139]. Belonging to the Flavivirus genus, which is part of the Flaviviridae family, Dengue virus (DENV) consists of four serotypes, namely DENV-1, -2, -3, and DENV-4, which are antigenically similar but genetically distinct. The nucleotide sequences of each of the four DENV serotypes differ by 25–35 bp, and each serotype has the potential to cause dengue fever. Among these four serotypes, DENV-4 is the most divergent, followed by DENV-2. DENV-1 and DENV-3 exhibit nearly equal levels of divergence. Dengue fever, which is a mild feverish illness, is caused by infection from any of the four serotypes. The probability of developing hemorrhagic fever/dengue shock syndrome (DHF/DSS), which results from significant bleeding and capillary plasma leakage, increases when multiple serotypes are present during the infection. Infection with a particular serotype provides long-lasting immunity to that particular serotype, but only temporary protection against the remaining three serotypes. Based on epidemiological studies, being infected with various serotypes subsequently could potentially lead to a more severe form of dengue fever. Dengue fever is now present in more than 100 countries worldwide in the tropics and subtropics. It occurs with 1 to 30% mortality and puts nearly 2.5 billion people at risk. Prevention of dengue fever is difficult due to overcrowding, rapid urbanization, increased travel, and the lack of effective measures to control vector mosquitoes. In addition, no dengue vaccine has yet been approved. Furthermore, clinical symptoms of dengue are similar to those of leptospira, rickettsia, chikungunya, and malaria. The importance of quick and reliable laboratory diagnostic tests cannot be overstated when it comes to confirming illnesses and prioritizing patient care. Specifically, for diagnosing the dengue virus, there are several conventional techniques available. These include serological testing, viral detection in cell culture, and reverse transcriptase PCR amplification. These established methods are essential for accurately detecting the presence of the virus and guiding appropriate medical interventions. Given the absence of effective vaccination or treatment options for this disease, the demand for a dependable diagnostic tool for swift and ongoing detection becomes crucial. Isolating the virus or identifying specific dengue antibodies can be a laborious and time-consuming process. The PCR method, although effective, necessitates advanced technical expertise and can be prone to laboratory contamination. Fortunately, biosensors offer a solution by overcoming all of these limitations [140].

In recent times, bioscience experts have been dedicated to developing optimal biosensors for identifying the Dengue virus. These biosensors are designed to achieve high levels of simplicity, specificity, sensitivity, and accuracy. The most commonly used biosensors include microfluidic, electrical, electrochemical, optical, and smartphone-based biosensors. These biosensors demonstrate both high accuracy and sophistication, allowing them to effectively differentiate all Dengue serotypes from other viruses by utilizing specific bio-receptors. Overall, these advancements in biosensor technology offer promising tools for accurate and reliable detection of the Dengue virus, enabling timely interventions and improved management of the disease. These sensors are very promising for clinical applications due to the aforementioned advantages and have the potential to solve many past challenges, such as attaining high accuracy and detection rate simultaneously. For instance, there was a report on a semi-disposable microfluidic biosensor that could detect the sequence of the 3-dengue serotype. By utilizing a liposomal encapsulated fluorescent reagent to enhance the signal, the sensor was able to detect 20 to 25 long-recording oligonucleotides with a limit of detection of 10 pmol L^{-1}. Recent reports have highlighted the application of a fluorescent safety sensor based on microfluidic dielectrophoresis (DEP) device for detecting the dengue virus. This approach was able to identify modest volumes of samples (15 μL) in just 5 min. In this particular test, the DEP force mechanism was employed to trap modified beads, while the Dengue virus was labeled with a fluorescent marker to facilitate detection. This method demonstrated notable advantages over previous approaches, requiring less time, cost, and sample volume.

Kwakye et al. conducted a significant investigation involving the development of a novel electrochemical microfluidic biosensor [141]. The biosensor included a reporter probe inside liposomes that contained a redox reagent, while the capture probe was linked to superparamagnetic particles. The hybridization procedure, followed by magnetic separation, led to the disintegration of liposomes and the subsequent liberation of $Fe\ (CN)_6^{3-/4-}$. This released compound was then accurately quantified using the method of amperometry. It is noteworthy that this particular microfluidic biosensor achieved an impressive limit of detection (LOD) of 0.01 μM. The utilization of microfluidic techniques offers substantial diagnostic potential in detecting the presence of the Dengue virus. Moreover, these techniques are commercially available and cost-effective, requiring only a small sample size of 3–10 μL of blood for analysis. Alternatively, employing kit-type detection devices that utilize rapid tests based on microfluidic dielectrophoresis chips is another viable option for swift virus diagnosis. This novel biosensor can identify the Dengue virus in less than 5 min. The multiplex kits allow for the simultaneous visual identification of Chikungunya, Zika, and Dengue viruses in about thirty minutes [142].

Other intriguing bio-devices that facilitate quick testing and promote the commercialization of portable biosensors for Dengue virus diagnostics include paper-based microfluidic kits. However, their utility has been limited due to modest sensitivity and a lack of fluid flow control. Agarose-based papers, on the other hand, may serve as a suitable material for this purpose.

Several commercially available kits have been developed for early Dengue virus diagnosis. These kits provide high assay rates and accuracy, making them widely applicable in clinical settings. In summary, the key advantages of microfluidics and diagnostic kits include their portability, acceptable accuracy, high assay rates, simple setup, and expertise in detecting the Dengue virus in non-blood specimens. Compact lab-on-a-chip (LOC) biosensors, diagnostic kits, and microfluidic devices are considered the most feasible options for clinical use due to their proficiency.

5.23 Smartphone-Based Methods

Smartphones are unique POC devices in nano bio sensing because of their popularity, mobility, small components, data collecting and transmission capacities, cost-effectiveness, and high-quality camera lenses. On the other hand, one of the essential features of biosensors in the commercialization process is their ease of use. In recent years, employing smartphones for this purpose has gained a lot of interest since it simplifies the detection process and eliminates the need for sophisticated devices. Ganguli and colleagues created a point-of-care system that uses a smartphone and a microfluidic platform based on real-time RT-LAMP (Reverse Transcription Loop-Mediated Isothermal Amplification) technology. It is a molecular approach for rapidly and efficiently amplifying particular RNA sequences). This configuration is effectively a LOC platform capable of detecting Zika, the Chikungunya, and Dengue viruses simultaneously [143]. Another solution that can help achieve low cost, acceptable performance, and accuracy is to combine the RT-LAMP approach with turning off unreliable signal reporting (QUASR) technology. As a result, smartphone-based biosensors are a very effective solution for low-cost, point-of-care techniques. The most important benefits of this approach short test time, are ease of use, and good performance [143].

5.24 Covid-19 Diagnosis

In December 2019, a new type strain of beta-coronavirus known as SARS-CoV-2 surfaced in Wuhan, China, resulting in the development of severe acute respiratory syndrome and ultimately resulting in a worldwide COVID-19 epidemic. The virus is highly transmissible and can be spread through respiratory droplets and aerosols. It can be transmitted by both symptomatic and asymptomatic individuals. As a result, the World Health Organization (WHO) advises taking certain precautions such as ensuring proper ventilation, avoiding crowded places, maintaining physical distancing, and wearing masks. The COVID-19 outbreak has had a profound impact on global health, especially the economic and social systems worldwide. This impact is primarily due to the implementation of social distancing measures and lockdowns. It is crucial to quickly detect and test for the virus to assess the risks associated with reopening schools, workplaces, and other social settings. However, the current diagnostic tests for the virus are centralized, requiring costly laboratory equipment and skilled technicians. Moreover, the testing process itself is time-consuming, which poses challenges in meeting the high demand for testing. The development of effective diagnostic platforms for COVID-19 requires a comprehensive understanding of the virus's biological and clinical characteristics. The coronavirus is a relatively large virus that features an envelope with spike proteins, enabling it to invade human cells by binding to ACE receptors. Patients infected with SARS-CoV-2 exhibit a range of symptoms, including dry cough, dyspnea, fever, loss of taste and/or smell, and muscular discomfort. To accurately detect and track the virus, it is essential to develop lab-on-chip applications that utilize relevant protein and nucleic acid biomarkers. These applications can facilitate the creation of sensitive, fast, and cost-effective diagnostic devices for accurate screening of individuals. There are three commonly sought-after biomarkers for viral diagnosis: viral proteins present on the viral envelope (detected through antigen detection test), antibodies to the virus detected through serological assays (detected through serological test), and viral genetic material in the form of DNA or RNA (detected through viral gene detection). These biomarkers, as shown in Fig. 5.9, play a crucial role in effectively monitoring the spread of the virus, properly isolating infected individuals, and providing timely treatment to patients. Reverse transcription polymerase chain reaction (RT-PCR) is considered the most reliable method for detecting viral genes due to its high sensitivity and signal amplification capabilities. Nonetheless, it has some limitations, such as being a time-consuming process taking 2–5 h, requiring specialized laboratory settings, and incurring significant expenses. Hence, it may not be suitable for mass population testing. In contrast, the molecular biology technique of reverse transcription loop-mediated isothermal amplification (RT-LAMP) is a relatively new approach that shows great promise. This method amplifies the virus at a single temperature and does not require specialized laboratory infrastructure, making it a more convenient option. Nevertheless, trained personnel are still necessary to carry out the processing steps, and it is not an ideal choice for a point-of-care (POC) platform. Henceforth, it is imperative to expedite the creation of point-of-care (POC) diagnostic mechanisms that exhibit qualities such as precision, promptness, and economic viability. These devices are essential for accurate screening of individuals and tracking the spread of the virus. By utilizing these technological advancements, it becomes possible to effectively segregate individuals who have contracted the virus and promptly deliver medical care, both of which are crucial in managing the COVID-19 outbreak. Various techniques are being used in different countries to detect SARSCoV-2 antibodies. One such technique is lateral flow immunoassays, which provide quick results and do not require specialized equipment. These tests are capable of identifying both IgM and IgG antibodies in whole blood, plasma, or serum samples. However, it is important to note that these tests cannot determine the precise quantities of antibodies present in the test subjects. On the other hand, ELISA tests, although taking longer to complete as they require separation of serum or plasma, can detect a broader range of antibodies. Microfluidic devices have the potential to detect SARS-CoV-2 antibodies cost-effectively and reliably. ELISA tests, which stands for Enzyme Linked Immunosorbent Assay tests, have been widely utilized during the current pandemic. The ELISA technique comprises four main variations: Direct, Indirect, Sandwich, and Competitive ELISA. These different types are illustrated in Fig. 5.10 [144]. In Direct ELISA, enzyme-linked antibodies directly bind to antigens that are immobilized on a surface, leading to the generation of a detectable signal. Indirect ELISA involves the initial binding of a primary antibody to the antigen, followed by the binding of a secondary antibody that is linked to an enzyme. In Sandwich ELISA, unlike Direct and Indirect ELISA, the antibody is immobilized on the surface (known as the capture antibody) rather than the antigen. The antigen then binds to this capture antibody. Subsequently, an enzyme-labeled antigen attaches to the captured antigen, initiating an enzyme–substrate reaction that produces a signal. The three aforementioned techniques can also be adapted to a fourth protocol called competitive ELISA. In competitive ELISA, there is an additional reference antigen that competes with the target antigen for binding to the antibodies. Hence, a weaker signal indicates a higher concentration of the target antigen. It is worth noting that ELISA tests typically require the separation of plasma or serum and generally have longer completion times, often spanning several hours. Impedance-based sensors and electrochemical-based sensors are two types of microfluidic devices that have been developed for detecting and quantifying antibodies. While these devices cannot directly analyze human blood samples, they are capable of accurately measuring antibody levels in

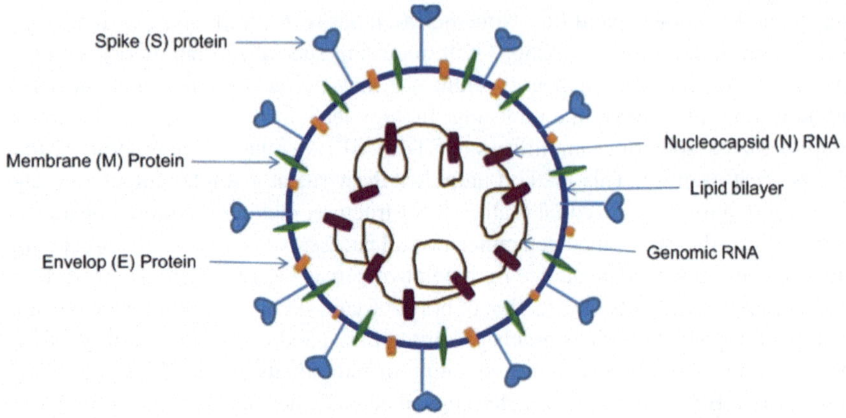

Fig. 5.9 The composition and biomarkers of the SARS-CoV-2 virus: visualizing viral RNA and key protein components [147]

human serum within a short timeframe of 30 min. However, for practical implementation in point-of-care (POC) settings, these devices need to be automated to handle the required incubation and washing steps. Additionally, researchers have made significant progress in developing optical detection sensors that offer an alternative solution. These sensors have the advantage of being compatible with small quantities of unprocessed whole blood, making them a promising option for POC testing [145]. Research has explored Localized Surface Plasmon Resonance (LSPR) and Surface Plasmon Resonance (SPR) as potential techniques for detecting target molecules. However, the creation of an economical and portable reader for these sensing methods continues to pose a challenge. When it comes to detecting and measuring SARS-CoV-2 antibodies, various approaches exist, each with its own distinct benefits and limitations [146, 147].

5.25 Electrochemical-Based Sensors

Electrochemical-based sensors are remarkable microfluidic devices that utilize the principles of sandwich-type electrochemistry to detect and measure antibodies in biological samples. They achieve this by monitoring the electrical response generated by a chemical reaction, which directly correlates with the concentration of the target molecules being sought. A notable example of such a device is a screen-printed microfluidic device developed by researchers, specifically designed to quantify IgG antibodies in serum samples. The device features electrodes coated with anti-mouse IgG, which are firmly attached using electropolymerized polypyrrole propylic acid (PPA) film, serving as a highly effective binding agent. The device used in the study involves a multi-step process for introducing reagents. The process involves using

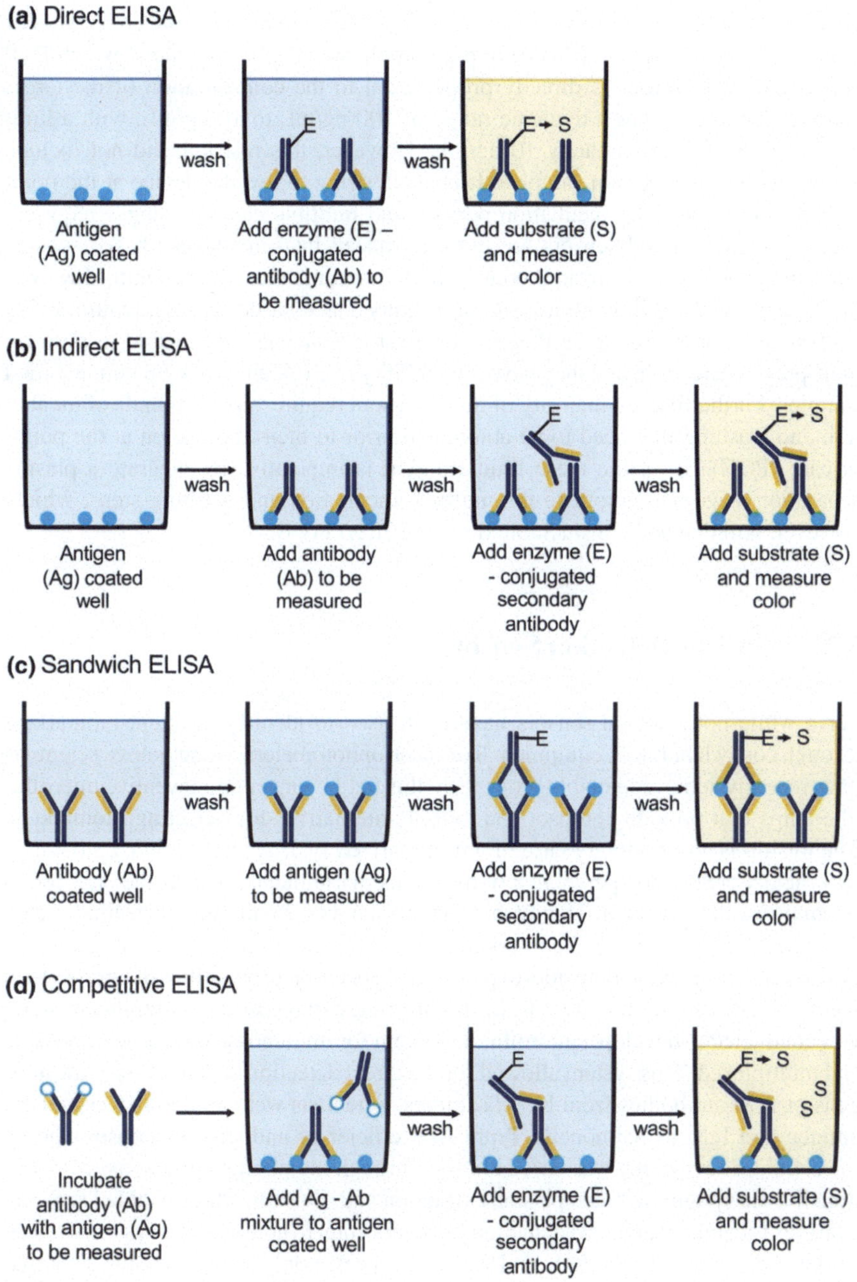

Fig. 5.10 Schematic presentation of basic types of ELISA (enzyme-linked immunosorbent assay): **a** direct, **b** indirect, **c** sandwich, **d** competitive; Ab antibody, Ag antigen, S substrate, E enzyme. Reprinted with permission from [148]

alkaline phosphatase-conjugated secondary antibodies to measure the conversion of p-aminophenyl phosphate (PAPP) to p-aminophenol (PAP). The rate at which PAP concentration increases is directly proportional to the concentration of the target analyte. The device has a dynamic range of 100 ng/mL to 10 ug/mL, with a limit of detection of approximately 10 ng/mL. However, this research did not include testing with human serum samples. It is challenging to use the device at the point of care due to the long incubation periods and multiple cleaning stages required. Electrochemical-based sensors are more advanced than impedance-based sensors and can efficiently measure antibodies in human serum in less than 30 min. However, they are not without limitations. Similar to impedance-based sensors, require bulky instrumentation for the quantification of electrical signals, are unable to perform multiplexed detection, and they have the inability to work directly with human blood samples. Furthermore, a majority of these sensors require various rounds of incubation and washing that need to be automated prior to their application at the point-of-care (POC). To tackle these limitations, it is imperative to integrate a plasma separation stage and automate the multiple incubation and washing steps, which, however, will demand considerable time and effort [149].

5.26 Optical Detection Sensors

For a while now, optical sensors have been used to identify different biomarkers through costly laboratory equipment like spectrophotometers. Nonetheless, scientists have recently achieved notable progress in this field by inventing inventive microfluidic chips that provide compact and mobile alternatives for detecting biomarkers. The mChip is an exemplary case of a miniaturized ELISA microfluidic chip that is specifically created to operate in low-resource environments, such as the ones found in countries like Rwanda. The mChip, which stands for a mobile microfluidic chip for immunoassay on protein biomarkers, utilizes optical detection technology, which utilizes LEDs and photodetectors to precisely detect minute quantities of unprocessed whole blood, making it a great fit for point-of-care diagnostics. A significant study was conducted to develop a microfluidic system for immunoassays that is disposable and multiplexed. This system allowed for the direct detection of disease-specific antigens or IgM antibodies from blood samples. Scientists were able to detect malaria antigen and IgM to Salmonella Typhi LPS efficiently and simultaneously using a microfluidic device that was based on flow-through membrane immunoassay. Additionally, the system included plasma extraction and division, which enabled separate analyses for antigen and IgM detection. It is important to note that this system showed a limit of detection of 20 ng/mL, and the assay time was only 30 min, making it comparable to traditional benchtop ELISA tests in terms of performance. In the realm of target molecule detection, localized surface plasmon resonance (LSPR) and surface plasmon resonance (SPR) have been scrutinized as potential sensing methods. One group of researchers, for instance, experimented with SPR phase imaging to identify IgG antibodies. To overcome the issue of temperature sensitivity in SPR, the team included temperature sensors and microheaters in their system. By harnessing the power of a monochromatic light source and a CCD device to evaluate the reflected light, the scientists were able to achieve precise IgG detection [150].

5.27 A Digital Microfluidic (DMF)

A newly introduced digital microfluidic (DMF) device was developed to evaluate the levels of human IgG antibodies in BSA. Yet, the practical implementation of these sensors in point-of-care settings faces challenges due to the need for manual sample preparation and the requirement of bulky optical instruments for data collection. As a solution, an alternative biosensor was created to detect human serum IgG antibodies specific to Helicobacter pylori. This innovative biosensor employed antigen-coated beads, a secondary antibody tailored to human IgG and coupled with an enzyme, along with 4-methylumbelliferyl phosphate (4-MUP) as the enzymatic substrate. It is worth noting that this biosensor achieved an assay time of 30 min and demonstrated an impressive detection limit of 0.17 U/ml. However, prior to usage, serum dilution remained necessary, introducing additional manual processing steps. The researchers successfully showed the design and performance of an integrated microfluidic sensor designed particularly for detecting dengue virus IgM and IgG antibodies in human serum. The fluorescent markers used for detection were anti-human IgM-R-PE antibodies and anti-human IgG-FITC antibodies. The assay time for this sensor was approximately 30 min, and it exhibited a reported detection limit of 21 pg. One notable advantage of this biochip is its ability to simultaneously detect both IgM and IgG in human serum. Despite the involvement of multiple washing and incubation steps in the setup, all of these processes have been automated and carried out on the chip itself, significantly enhancing the feasibility of utilizing this sensor at point-of-care (POC) settings. The utilization of chemiluminescent methodologies in conjunction with microfluidic devices offers a user-friendly and rapid approach for protein and antibody identification. These tests typically exhibit a direct correlation between the concentration of antibodies and the chemiluminescent output, providing valuable insights into the presence and levels of specific proteins and antibodies [151].

5.28 Progress in Paper-Based Biosensors for Use in On-Site Evaluations

Paper-based biosensors (PBBs) have received a lot of interest in the development of Point-of-Care (POC) devices because of their simplicity, low cost, and user-friendliness. The use of nanoparticles (NPs) as labels is critical in the design and manufacture of PBB strips. The selection of NPs and the associated detection mechanism have a direct impact on the performance of these devices. Figure 5.11

shows three types of testing: dipstick tests, lateral flow assays (LFAs), and microfluidic biosensors (μPADs) [152]. Microfluidics, a multidisciplinary technique that studies the flow of liquids via micrometer-sized channels, has enormous promise for improving current diagnostic instruments, making them more precise, efficient, and cost-effective. Microfluidic paper-based analytical devices (μPADs), or paper-based microfluidics, are diagnostic systems made of patterned paper substrates. Capillary action propels a tiny amount of fluid through the multi-channel designs embedded in the paper substrate. Figure 5.11 shows an example of such a gadget. Paper-based microfluidics are ideal for multiplex testing of various biological samples, such as whole blood, serum, urine, and nose swabs. These μPADs provide many detection modalities, including colorimetric, electrochemical, chemiluminescence, and electrochemiluminescence. Furthermore, they enable automated sample processing, which is consistent with the ASSURED device's objective of being affordable, sensitive, specific, user-friendly, rapid, equipment-free, and deliverable to end users. Paper is a good material for clinical analysis due to its biological compatibility and ability to absorb reagents efficiently. Its naturally white backdrop offers an ideal contrast for color-based detection systems. Furthermore, paper is lightweight, inexpensive, and widely available, making it an environmentally benign option for building portable gadgets. Several 3D paper-based microfluidic biosensors have been created for the point-of-care detection of proteins or nucleic acids using fluorescence or colorimetric detection methods. Integrated devices that combine μPADs with camera phones or other imaging devices to quantify colorimetric findings and communicate data to qualified clinicians have been applied effectively. These applications constitute the first steps toward using telemedicine for cost-effective human health monitoring, which is especially important in underdeveloped countries [153].

5.29 Paper/Colorimetric Detection

A microfluidic device was developed using paper and polymer materials to measure Hepatitis B surface Antigen (HBsAg) and Immunoglobulin G (IgG). The device utilized BCIP/NBT-conjugated secondary antibodies and alkaline phosphatase (ALP), enabling detection limits of 1.3 ng/mL for HBsAg and 1.6 ng/mL for IgG. However, it had an assay time of approximately an hour and could not measure IgG in human serum samples. Recent advancement in immunoassays involves a paper-based method that utilizes bioluminescence resonance energy transfer (BRET) to detect antibodies in blood serum. This innovative device incorporates antibody binding and signal generation into a protein switch called LUMABS, allowing simultaneous detection of three antibodies in whole blood or plasma. It is user-friendly and offers a quick assay time of just 20 min. Rapid and cost-effective antigen detection assays that utilize specific antibodies to identify viral antigens have become important diagnostic tools. These tests are suitable for point-of-care (POC) settings and have played a significant role in the widespread detection of COVID-19. However, the limited availability of these assays on the market and the lack of published research poses

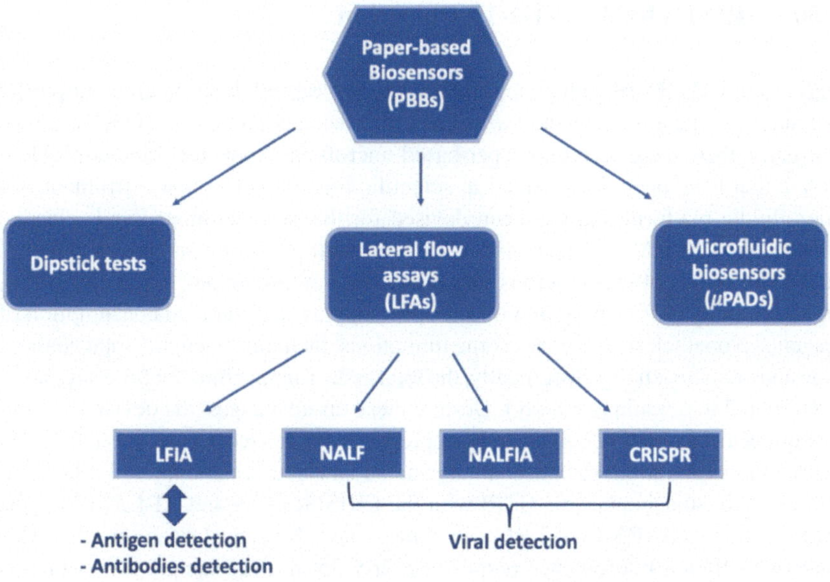

Fig. 5.11 PBB categorization scheme schematic. Abbreviations: CRISPR stands for clustered regularly interspaced short palindromic repeats; LFIA stands for lateral flow immunoassay; NALF for nucleic acid lateral flow; and NALFIA for nucleic acid lateral flow immunoassay. Reprinted with permission from MDPI journal [152]

challenges. Serological assays are used to detect antibodies produced by infected individuals at different stages of the disease. The most commonly utilized technique is the biochemical enzyme-linked immunosorbent assay (ELISA). However, ELISA necessitates trained personnel and specialized laboratories, which makes it impractical for point-of-care (POC) applications. Affinity-based biosensors (ABBs) and paper-based biosensors (PBBs) are emerging as viable alternatives to traditional diagnostic methods. These biosensors offer self-contained diagnostic tools with rapid response times and satisfactory sensitivity. By adhering to the ASSURED criteria guidelines for POC testing, they aim to provide same-day diagnosis and facilitate prompt decision-making. Understanding these concepts provides valuable insights into the advancements made in microfluidic devices, paper-based tests, antigen and serological examinations, challenges encountered in detecting COVID-19 antibodies, and the potential of paper-based and affinity-based biosensors as essential analytical instruments [154].

5.30 MPADs for COVID-19 Detection

During the COVID-19 pandemic, scientists have been working on creating μPAD techniques to meet the demand for fast and affordable point-of-care (POC) options. Currently, there have been two paper-based microfluidic tools for detecting SARS-CoV-2 that have been documented in scientific literature. The first instrument is a microfluidic platform that has been devised for the semi-automated and selective identification of IgM and IgG antibodies from multiple serum samples simultaneously. This device offers numerous advantages, such as throughput, specificity (91%), high sensitivity (95%), detection of multiple analytes, and minimal consumption of reagents. Nonetheless, it also has some limitations, including a lengthy total analysis time and cross-reactivity. Specifically, the total assay running time for 50 assays takes 5.5 h (equal to 6.6 min per assay). Despite these disadvantages, the device's overall strong performance renders it highly appropriate for low-cost mass screenings. In the second exploration, scientists amalgamated an electrokinetic microfluidic technique called isotachophoresis (ITP) with the CRISPR assay and RT-LAMP for the identification of SARS-CoV-2 RNA in clinical nasopharyngeal swab samples. This approach offers a fluorescence output in nearly 35 min per sample row, which is a noteworthy upgrade compared to current CRISPR-based methods for COVID-19 detection. Nevertheless, the RT-LAMP stage was executed outside the chip in this analysis, and further exploration is mandatory to merge all assay steps into a portable microfluidic system. These innovative contraptions hold great potential in terms of swiftly and economically addressing Point of Care demands amidst the COVID-19 crisis [152].

5.31 Conclusion

In conclusion, this chapter has extensively explored the integration of biosensors and microfluidics and its diverse applications. It began by discussing the concept of on-chip biosensors, highlighting their compact and integrated nature. The focus then shifted to continuous microfluidic-based biosensors, specifically examining droplet and digital microfluidic (DMF)-based biosensors. The chapter further explored recent advancements in microfluidic operating systems, with a particular emphasis on the emerging field of paper-based chips. These platforms offer cost-effective and portable solutions for a wide range of applications. The integration of microfluidics and biosensors for point-of-care (POC) diagnostics was thoroughly investigated, with notable examples such as glucose biosensors, cholesterol/lipid biosensors, urea sensors, and optofluidics being discussed. Furthermore, the chapter delved into the application of integrating microfluidics and biosensors in virus detection. The diagnosis of the Dengue virus was highlighted as a specific example, showcasing the innovative approach of smartphone-based methods. Additionally, the chapter explored the relevance of this integration in the context of COVID-19 detection, presenting various techniques including electrochemical-based optical detection sensors, digital microfluidics, paper/colorimetric detection, paper-based biosensors (PBBs), and microfluidic paper-based analytical devices (μPADs) for COVID-19 detection. Overall, this chapter has emphasized the significant progress and potential that arises from the fusion of biosensors and microfluidics in various applications, encompassing POC diagnostics and virus detection. These collaborative technologies hold great promise in enhancing diagnostic capabilities, improving portability, and enabling rapid and sensitive detection in diverse healthcare settings. The findings from this chapter underscore the importance of further research and development in this field, as ongoing innovation will continue to revolutionize the realm of biosensing and microfluidics.

References

1. Sabry, F.: Lab On A Chip: Low-Cost Point-of-Care Devices for Human Diseases Diagnosis, Possibly Making Laboratories Dispensable, vol. 9. One Billion Knowledgeable (2022)
2. Mishra, R., Bhattacharyya, T.K.: MEMS-Based Transdermal Drug Delivery. CRC Press (2023)
3. Pandey, Y., Singh, S.P.: Recent advances in bio-MEMS and future possibilities: an overview. J. Inst. Eng. (India): Ser. B **104**(6), 1377–1388 (2023)
4. Giouroudi, I., Kokkinis, G.J.N.: Recent advances in magnetic microfluidic biosensors. **7**(7), 171 (2017)
5. Mikaeeli Kangarshahi, B., Naghib, S.M., Rabiee, N.: DNA/RNA-based electrochemical nanobiosensors for early detection of cancers. Crit. Rev. Clin. Lab. Sci. 1–23
6. Naghib, S.M., et al.: Synthesis, characterization and electrochemical evaluation of polyvinylalchol/graphene oxide/silver nanocomposites for glucose biosensing application. Int. J. Electrochem. Sci. **13**(1), 1013–1026 (2018)
7. Salahandish, R., et al.: A novel graphene-grafted gold nanoparticles composite for highly sensitive electrochemical biosensing. IEEE Sens. J. **18**(6), 2513–2519 (2018)
8. Salahandish, R., et al.: Nano-biosensor for highly sensitive detection of HER2 positive breast cancer. Biosens. Bioelectron. **117**, 104–111 (2018)
9. Salahandish, R., et al.: Label-free ultrasensitive detection of breast cancer miRNA-21 biomarker employing electrochemical nano-genosensor based on sandwiched AgNPs in PANI and N-doped graphene. Biosens. Bioelectron. **120**, 129–136 (2018)
10. Naghib, S.M., Katebi, S.M., Ghorbanzade, S.: Electrochemical Biosensors in Practice: Materials and Methods. Bentham Science Publishers (2023)
11. Sadana, A., Sadana, N.: Handbook of Biosensors and Biosensor Kinetics. Elsevier (2010)
12. Ozkan, S.A., Uslu, B., Sezgintürk, M.K.: Biosensors: Fundamentals, Emerging Technologies, and Applications. CRC Press (2022)
13. Naghib, S.M., et al.: Investigation of a biosensor based on phenylalanine dehydrogenase immobilized on a polymer-blend film for phenylketonuria diagnosis. Electroanalysis **24**(2), 407–417 (2012)
14. Sartipzadeh, O., et al.: Controllable size and form of droplets in microfluidic-assisted devices: effects of channel geometry and fluid velocity on droplet size. Mater. Sci. Eng. C **109**, 110606 (2020)
15. Sartipzadeh, O., et al.: Microfluidic-assisted synthesis and modelling of monodispersed magnetic nanocomposites for biomedical applications. **9**(1), 1397–1407 (2020)

16. Sartipzadeh, O., et al.: Microfluidic-assisted synthesis and modeling of stimuli-responsive monodispersed chitosan microgels for drug delivery applications. Sci. Rep. **12**(1), 8382 (2022)
17. Rahmanian, M., et al.: A micropillar array-based microfluidic chip for label-free separation of circulating tumor cells: the best micropillar geometry? J. Adv. Res. **47**, 105–121 (2023)
18. Tian, W.-C., Finehout, E.: Introduction to microfluidics. In: Microfluidics for Biological Applications, pp. 1–34. Springer (2008)
19. Safavieh, R., Tamayol, A., Juncker, D.: Serpentine and leading-edge capillary pumps for microfluidic capillary systems. Microfluid. Nanofluid. **18**(3), 357–366 (2015)
20. Rahul, R., et al.: A mould-free soft-lithography approach for rapid, low-cost and bulk fabrication of microfluidic chips using photopolymer sheets. Microfluid. Nanofluid. **27**(11), 78 (2023)
21. Manoj, H., et al.: Microneedles: current trends and applications. In: Microfluidics and Bio-MEMS, pp. 275–342. Jenny Stanford Publishing (2020)
22. Nikoleli, G.-P., et al.: Biosensors Based on Microfluidic Devices Lab-on-a-Chip and Microfluidic Technology, pp. 375–394 (2018)
23. Luka, G., et al.: Microfluidics integrated biosensors: a leading technology towards lab-on-a-chip and sensing applications. **15**(12), 30011–30031 (2015)
24. Luka, G., et al.: Microfluidics integrated biosensors: a leading technology towards lab-on-a-chip and sensing applications. Sensors **15**(12), 30011–30031 (2015)
25. Nikoleli, G.-P., et al.: Chapter 13—biosensors based on microfluidic devices lab-on-a-chip and microfluidic technology. In: Nikolelis, D.P., Nikoleli, G.-P. (eds.) Nanotechnology and Biosensors, pp. 375–394. Elsevier (2018)
26. Liao, Z., et al.: Recent advances in microfluidic chip integrated electronic biosensors for multiplexed detection. Biosens. Bioelectron. **121**, 272–280 (2018)
27. Goode, J.A., Rushworth, J.V.H., Millner, P.A.: Biosensor regeneration: a review of common techniques and outcomes. Langmuir **31**(23), 6267–6276 (2015)
28. Sadeghi, M., et al.: A high-performance electrochemical aptasensor based on graphene-decorated rhodium nanoparticles to detect HER2-ECD oncomarker in liquid biopsy. Sci. Rep. **12**(1) (2022)
29. Naghib, S.M., Zare, Y., Rhee, K.Y.: A facile and simple approach to synthesis and characterization of methacrylated graphene oxide nanostructured polyaniline nanocomposites. Nanotechnol. Rev. **9**(1), 53–60 (2020)
30. Mohammadpour, Z., Naghib, S.M.: Chapter 15—smart nanosensors for intelligent packaging. In: Thomas, S., et al., (eds.) Nanosensors for Smart Manufacturing, pp. 323–346. Elsevier (2021)
31. Naghib, S.M.: Two-dimensional functionalised methacrylated graphene oxide nanosheets as simple and inexpensive electrodes for biosensing applications. Micro Nano Lett. **14**(4), 462–465 (2019)
32. Shahbazi, N., Zare-Dorabei, R., Naghib, S.M.: Multifunctional nanoparticles as optical biosensing probe for breast cancer detection: a review. Mater. Sci. Eng. C **127**, 112249 (2021)
33. Mohanty, S.P., Kougianos, E.: Biosensors: a tutorial review. IEEE Potentials **25**(2), 35–40 (2006)
34. Matini, A., Naghib, S.M., Mozafari, M.R.: Quantum Dots in Cancer Theranostics: A Thorough Review of Recent Advancements in Bioimaging, Tracking, and Therapy across Various Cancer Types. Current Pharmaceutical Biotechnology (2024)
35. Shanehband, N., Naghib, S.M.: Microfluidics-assisted tumor cell separation approaches for clinical applications: an overview on emerging devices. Comb. Chem. High Throughput Screen (2024)
36. Shanehband, N., Naghib, S.M.: Recent advances in nano/microfluidics-based cell isolation techniques for cancer diagnosis and treatments. Biochimie **220**, 122–143 (2024)
37. Shahbazi, N., Zare-Dorabei, R., Naghib, S.M.: Design of a ratiometric plasmonic biosensor for herceptin detection in HER2-positive breast cancer. ACS Biomater. Sci. Eng. **8**(2), 871–879 (2022)

38. Sadeghi, M., et al.: A highly sensitive nanobiosensor based on aptamer-conjugated graphene-decorated rhodium nanoparticles for detection of HER2-positive circulating tumor cells. **11**(1), 793–810 (2022)
39. Rahimzadeh, Z., et al.: A rapid nanobiosensing platform based on herceptin-conjugated graphene for ultrasensitive detection of circulating tumor cells in early breast cancer. **10**(1), 744–753 (2021)
40. Naghib, S.M., Zare, Y., Rhee, K.Y.: A facile and simple approach to synthesis and characterization of methacrylated graphene oxide nanostructured polyaniline nanocomposites. **9**(1), 53–60 (2020)
41. Naghib, S.M., et al.: A highly sensitive biosensor based on methacrylated graphene oxide-grafted polyaniline for ascorbic acid determination. **9**(1), 760–767 (2020)
42. Naghib, S.M.: Localized Micro/Nanocarriers for Programmed and On-Demand Controlled Drug Release. Bentham Science Publishers (2022)
43. Arora, N.: Recent advances in biosensors technology: a review. Octa J. Biosci. **1**(2) (2013)
44. Garipcan, B., Çağlayan, M.O., Demirel, G.: New generation biosensors based on ellipsometry. New Perspectives in Biosensors Technology and Applications, pp. 197–214 (2011)
45. Kangarshahi, B.M., Naghib, S.M.: Nanogenosensors based on aptamers and peptides for bioelectrochemical cancer detection: an overview of recent advances in emerging materials and technologies. Discov. Appl. Sci. **6**(2), 47 (2024)
46. Naresh, V., Lee, N.: A review on biosensors and recent development of nanostructured materials-enabled biosensors. Sensors **21** (2021). https://doi.org/10.3390/s21041109
47. Korotkaya, E.V.: Biosensors: design, classification, and applications in the food industry. Foods Raw Mater. **2**(2), 161–171 (2014)
48. Karunakaran, C., Rajkumar, R., Bhargava, K.: Chapter 1—introduction to biosensors. In: Karunakaran, C., Bhargava, K., Benjamin, R., (eds.) Biosensors and Bioelectronics, pp. 1–68. Elsevier (2015)
49. Alhadrami, H.A.: Biosensors: classifications, medical applications, and future prospective. Biotechnol. Appl. Biochem. **65**(3), 497–508 (2018)
50. Minopoli, A., et al.: Ultrasensitive antibody-aptamer plasmonic biosensor for malaria biomarker detection in whole blood. Nat. Commun. **11**(1), 6134 (2020)
51. Yakoh, A., et al.: Paper-based electrochemical biosensor for diagnosing COVID-19: detection of SARS-CoV-2 antibodies and antigen. Biosens. Bioelectron. **176**, 112912 (2021)
52. Zhang, L., et al.: Antibody-gold nanoparticle bioconjugates for biosensors: synthesis, characterization and selected applications. Biosens. Bioelectron. **165**, 112370 (2020)
53. Sadique, M.A., et al.: Highly sensitive electrochemical immunosensor platforms for dual detection of SARS-CoV-2 antigen and antibody based on gold nanoparticle functionalized graphene oxide nanocomposites. ACS Appl. Bio Mater. **5**(5), 2421–2430 (2022)
54. Kondzior, M., Grabowska, I.: Antibody-electroactive probe conjugates based electrochemical immunosensors. Sensors **20** (2020). https://doi.org/10.3390/s20072014
55. Naresh, V., Lee, N.J.S.: A review on biosensors and recent development of nanostructured materials-enabled biosensors. **21**(4), 1109 (2021)
56. Korotkaya, E.J.F., Materials R.: Biosensors: design, classification, and applications in the food industry **2**(2) (2014)
57. Cavalcante, F.T.T., et al.: Designing of nanomaterials-based enzymatic biosensors: synthesis, properties, and applications. Electrochem **2**, 149–184 (2021). https://doi.org/10.3390/electrochem2010012
58. Ahangari, H., et al.: Latest trends for biogenic amines detection in foods: enzymatic biosensors and nanozymes applications. Trends Food Sci. Technol. **112**, 75–87 (2021)
59. Ramanavicius, S., Ramanavicius, A.: Charge transfer and biocompatibility aspects in conducting polymer-based enzymatic biosensors and biofuel cells. Nanomaterials **11** (2021). https://doi.org/10.3390/nano11020371
60. Mustafa, F., Andreescu, S.: Paper-based enzyme biosensor for one-step detection of hypoxanthine in fresh and degraded fish. ACS Sensors **5**(12), 4092–4100 (2020)

61. López Marzo, A.M., Mayorga-Martinez, C.C., Pumera, M.: 3D-printed graphene direct electron transfer enzyme biosensors. Biosens. Bioelectron. **151**, 111980 (2020)
62. Inshyna, N.M., et al.: Biosensors: design, classification and application. J. Nano- Electron. Phys. **12**(3) (2020)
63. Sempionatto, J.R., et al.: Epidermal enzymatic biosensors for sweat vitamin c: toward personalized nutrition. ACS Sens. **5**(6), 1804–1813 (2020)
64. Wu, L., et al.: Graphdiyne: a new promising member of 2D all-carbon nanomaterial as robust electrochemical enzyme biosensor platform. Carbon **156**, 568–575 (2020)
65. Jirakunakorn, R., et al.: Uric acid enzyme biosensor based on a screen-printed electrode coated with Prussian blue and modified with chitosan-graphene composite cryogel. Microchem. J. **154**, 104624 (2020)
66. Alvarado-Ramírez, L., et al.: Enzyme (single and multiple) and nanozyme biosensors: recent developments and their novel applications in the water-food-health nexus. Biosensors. **11** (2021). https://doi.org/10.3390/bios11110410
67. Kavita, V.: DNA biosensors-a review. J. Bioeng. Biomed. Sci. **7**(2), 222 (2017)
68. Du, Y., Dong, S.: Nucleic acid biosensors: recent advances and perspectives. Anal. Chem. **89**(1), 189–215 (2017)
69. Bialy, R.M., et al.: Functional nucleic acid biosensors utilizing rolling circle amplification. Chem. Soc. Rev. **51**(21), 9009–9067 (2022)
70. Chen, Y., et al.: Nucleic acid amplification free biosensors for pathogen detection. Biosens. Bioelectron. **153**, 112049 (2020)
71. Kaur, H., Bhosale, A., Shrivastav, S.: Biosensors: classification, fundamental characterization and new trends: a review. Int. J. Health Sci. Res. **8**(6), 315–333 (2018)
72. Kavita, V.J.J.B.B.S.: DNA biosensors-a review. **7**(2), 222 (2017)
73. Qin, Y., Li, Y., Hu, Y.: Emerging argonaute-based nucleic acid biosensors. Trends Biotechnol. **40**(8), 910–914 (2022)
74. Hashem, A., et al.: Nucleic acid-based electrochemical biosensors for rapid clinical diagnosis: advances, challenges, and opportunities. Crit. Rev. Clin. Lab. Sci. **59**(3), 156–177 (2022)
75. Zhang, Z., et al.: Development of nucleic-acid-based electrochemical biosensors for clinical applications. Angew. Chem. **134**(50), e202212496 (2022)
76. Gupta, N., et al.: Cell-based biosensors: recent trends, challenges and future perspectives. Biosens. Bioelectron. **141**, 111435 (2019)
77. Wei, X., et al.: Recent advances in electrochemical cell-based biosensors for food analysis: strategies for sensor construction. Biosens. Bioelectron. **248**, 115947 (2024)
78. Varshney, A., et al.: Microbial fuel cell-based biosensors and applications. Appl. Biochem. Biotechnol. **195**(5), 3508–3531 (2023)
79. Touhami, A.: Biosensors and Nanobiosensors: Design and Applications (2014)
80. Ronkainen, N.J., Halsall, H.B., Heineman, W.R.: Electrochemical biosensors. Chem. Soc. Rev. **39**(5), 1747–1763 (2010)
81. Wu, J., et al.: Device integration of electrochemical biosensors. Nat. Rev. Bioeng. **1**(5), 346–360 (2023)
82. Luka, G., et al.: Microfluidics integrated biosensors: a leading technology towards lab-on-a-chip and sensing applications. Sensors **15**, 30011–30031 (2015). https://doi.org/10.3390/s151229783
83. Borisov, S.M., Wolfbeis, O.S.: Optical biosensors. Chem. Rev. **108**(2), 423–461 (2008)
84. Cooper, M.A.: Optical biosensors in drug discovery. Nat. Rev. Drug Discovery **1**(7), 515–528 (2002)
85. Chen, C., Wang, J.: Optical biosensors: an exhaustive and comprehensive review. Analyst **145**(5), 1605–1628 (2020)
86. Pohanka, M.: The piezoelectric biosensors: principles and applications, a review. Int. J. Electrochem. Sci. **12**(1), 496–506 (2017)
87. Skládal, P.: Piezoelectric biosensors. TrAC, Trends Anal. Chem. **79**, 127–133 (2016)
88. Pohanka, M.: Overview of piezoelectric biosensors, immunosensors and DNA sensors and their applications. Materials **11** (2018). https://doi.org/10.3390/ma11030448

89. Skládal, P.: Piezoelectric biosensors: shedding light on principles and applications. Microchim. Acta **191**(4), 184 (2024)
90. Ramasamy, M.S., Bhaskar, R., Han, S.S.: Piezoelectric biosensors and nanomaterials-based therapeutics for coronavirus and other viruses: a mini-review. Curr. Top. Med. Chem. **23**(2), 115–127 (2023)
91. Wasilewski, T., et al.: Evaluation of linkers' influence on peptide-based piezoelectric biosensors' sensitivity to aldehydes in the gas phase. Int. J. Mol. Sci. **24** (2023). https://doi.org/10.3390/ijms241310610
92. Swain, J., et al.: Role of Piezoelectric Biosensors, in Point-of-Care Biosensors for Infectious Diseases, pp 129–145 (2023)
93. Kaur, A., et al.: Piezoelectric biosensors in healthcare. In: Patra, S., Kundu, D., Gogoi, M. (eds.) Enzyme-based Biosensors: Recent Advances and Applications in Healthcare, pp. 255–271. Springer Nature Singapore, Singapore (2023)
94. Forinová, M., et al.: A comparative assessment of a piezoelectric biosensor based on a new antifouling nanolayer and cultivation methods: enhancing S. aureus detection in fresh dairy products. Curr. Res. Biotechnol. **6**, 100166 (2023)
95. Almarri, N., et al.: Self-powered piezoelectric biosensing harvester for intracardiac monitoring. In: 2023 IEEE Biomedical Circuits and Systems Conference (BioCAS) (2023)
96. Karachaliou, C.-E., Livaniou, E.: Immunosensors for autoimmune-disease-related biomarkers: a literature review. Sensors **23** (2023). https://doi.org/10.3390/s23156770
97. Ramanathan, K., Danielsson, B.: Principles and applications of thermal biosensors. Biosens. Bioelectron. **16**(6), 417–423 (2001)
98. Petrini, G., et al.: Is a quantum biosensing revolution approaching? Perspectives in NV-assisted current and thermal biosensing in living cells. Adv. Quantum Technol. **3**(12), 2000066 (2020)
99. Xie, B., Danielsson, B.: Thermal biosensor and microbiosensor techniques. Handb. Biosens. Biochips **2**, 1–19 (2007)
100. Chen, S., Shamsi, M.H.: Biosensors-on-chip: a topical review. J. Micromech. Microeng. **27**(8), 083001 (2017)
101. Karaboğa, M.N.S., Sezgintürk, M.K.: Microfluidic based biosensors and applications. In: Biosensors, pp. 261–287. CRC Press (2022)
102. Patil, P.D., et al.: Microfluidic based continuous enzyme immobilization: a comprehensive review. Int. J. Biol. Macromol. **253**, 127358 (2023)
103. Song, Y., et al.: Recent progress in microfluidics-based biosensing. Anal. Chem. **91**(1), 388–404 (2019)
104. Choi, S., et al.: Microfluidic-based biosensors toward point-of-care detection of nucleic acids and proteins. Microfluid. Nanofluid. **10**(2), 231–247 (2011)
105. Li, C., et al.: A power-free deposited microbead plug-based microfluidic chip for whole-blood immunoassay. Microfluid. Nanofluid. **12**(5), 829–834 (2012)
106. Phurimsak, C., Tarn, M.D., Pamme, N.: Magnetic particle plug-based assays for biomarker analysis. Micromachines **7**, (2016). https://doi.org/10.3390/mi7050077
107. Itoh, D., et al.: Droplet-based microfluidic sensing system for rapid fish freshness determination. Sens. Actuators, B Chem. **171–172**, 619–626 (2012)
108. Rivet, C., et al.: Microfluidics for medical diagnostics and biosensors. Chem. Eng. Sci. **66**(7), 1490–1507 (2011)
109. Jiang, C., et al.: A cooperative multiagent reinforcement learning framework for droplet routing in digital microfluidic biochips. IEEE Trans. Comput. Aided Des. Integr. Circuits Syst. **42**(9), 3007–3020 (2023)
110. Rehman, A.U., et al.: Size-based sorting of emulsion droplets in microfluidic channels patterned with laser-ablated guiding tracks. Anal. Chem. **92**(3), 2597–2604 (2020)
111. Huang, S., et al.: Digital microfluidics for the detection of selected inorganic ions in aerosols **20**(5), 1281 (2020)
112. Samiei, E., et al.: Integration of biosensors into digital microfluidics: impact of hydrophilic surface of biosensors on droplet manipulation. Biosens. Bioelectron. **81**, 480–486 (2016)

113. Samiei, E., Tabrizian, M., Hoorfar, M.: A review of digital microfluidics as portable platforms for lab-on a-chip applications. Lab Chip **16**(13), 2376–2396 (2016)
114. Tao, X., Chakrabarty, K.: Functional testing of digital microfluidic biochips. In: 2007 IEEE International Test Conference (2007)
115. Choi, K., et al.: Digital microfluidics. Annu. Rev. Anal. Chem. **5**, 413–440 (2012)
116. Kulkarni, M.B., Ayachit, N.H., Aminabhavi, T.M.: Biosensors and microfluidic biosensors: from fabrication to application. Biosensors **12** (2022). https://doi.org/10.3390/bios12070543
117. Lee, V.B.C., et al.: Trends in paper-based electrochemical biosensors: from design to application. Anal. Sci. **34**(1), 7–18 (2018)
118. Liu, B., et al.: Paper-based electrochemical biosensors: from test strips to paper-based microfluidics. Electroanalysis **26**(6), 1214–1223 (2014)
119. Ge, X., et al.: Nanomaterial-enhanced paper-based biosensors. TrAC, Trends Anal. Chem. **58**, 31–39 (2014)
120. Ratajczak, K., Stobiecka, M.: High-performance modified cellulose paper-based biosensors for medical diagnostics and early cancer screening: a concise review. Carbohyd. Polym. **229**, 115463 (2020)
121. Loo, J.F.C., et al.: Integrated printed microfluidic biosensors. Trends Biotechnol. **37**(10), 1104–1120 (2019)
122. Kumar, S., et al.: Microfluidic-integrated biosensors: prospects for point-of-care diagnostics. Biotechnol. J. **8**(11), 1267–1279 (2013)
123. Pandey, C.M., et al.: Microfluidics based point-of-care diagnostics. Biotechnol. J. **13**(1), 1700047 (2018)
124. Agustini, D., Bergamini, M.F., Marcolino-Junior, L.H.: Tear glucose detection combining microfluidic thread based device, amperometric biosensor and microflow injection analysis. Biosens. Bioelectron. **98**, 161–167 (2017)
125. Liu, S., Su, W., Ding, X.: A review on microfluidic paper-based analytical devices for glucose detection. Sensors **16** (2016). https://doi.org/10.3390/s16122086
126. Sanati, A., et al.: Recent advancement in electrode materials and fabrication, microfluidic designs, and self-powered systems for wearable non-invasive electrochemical glucose monitoring. Appl. Mater. Today **26**, 101350 (2022)
127. Adeel, M., et al.: Glucose detection devices and methods based on metal-organic frameworks and related materials. Adv. Func. Mater. **31**(52), 2106023 (2021)
128. Fu, L.-M., Wang, Y.-N.: Detection methods and applications of microfluidic paper-based analytical devices. TrAC, Trends Anal. Chem. **107**, 196–211 (2018)
129. Kumar, S., et al.: Microfluidic-integrated biosensors: prospects for point-of-care diagnostics. **8**(11), 1267–1279 (2013)
130. Azahar Ali, M., et al.: Nanostructured anatase-titanium dioxide based platform for application to microfluidics cholesterol biosensor. Appl. Phys. Lett. **101**(8), 084105 (2012)
131. Ruecha, N., Siangproh, W., Chailapakul, O.: A fast and highly sensitive detection of cholesterol using polymer microfluidic devices and amperometric system. Talanta **84**(5), 1323–1328 (2011)
132. Kaur, G., Tomar, M., Gupta, V.: A simple paper based microfluidic electrochemical biosensor for point-of-care cholesterol diagnostics. Phys. Status Solidi (A) **214**(12), 1700468 (2017)
133. Pundir, C.S., Jakhar, S., Narwal, V.: Determination of urea with special emphasis on biosensors: a review. Biosens. Bioelectron. **123**, 36–50 (2019)
134. Chou, J.-C., et al.: The analysis of the urea biosensors using different sensing matrices via wireless measurement system & microfluidic measurement system. Sensors **19** (2019). https://doi.org/10.3390/s19133004
135. Chou, J.-C., et al.: The analysis of the urea biosensors using different sensing matrices via wireless measurement system and microfluidic measurement system. **19**(13), 3004 (2019)
136. Tang, J., Qiu, G., Wang, J.: Recent development of optofluidics for imaging and sensing applications. Chemosensors **10** (2022). https://doi.org/10.3390/chemosensors10010015
137. Monat, C., Domachuk, P., Eggleton, B.J.: Integrated optofluidics: a new river of light. Nat. Photonics **1**(2), 106–114 (2007)

138. Zhang, B., et al.: Diagnosing dengue virus infection: rapid tests and the role of micro/nanotechnologies. Nanomed.: Nanotechnol., Biol. Med. **11**(7), 1745–1761 (2015)
139. Zaytseva, N.V., Montagna, R.A., Baeumner, A.J.: Microfluidic biosensor for the serotype-specific detection of dengue virus RNA. Anal. Chem. **77**(23), 7520–7527 (2005)
140. Parkash, O., Shueb, R.H.: Diagnosis of dengue infection using conventional and biosensor based techniques. Viruses **7**, 5410–5427 (2015). https://doi.org/10.3390/v7102877
141. Kwakye, S., Goral, V.N., Baeumner, A.J.: Electrochemical microfluidic biosensor for nucleic acid detection with integrated minipotentiostat. Biosens. Bioelectron. **21**(12), 2217–2223 (2006)
142. Eivazzadeh-Keihan, R., et al.: Dengue virus: a review on advances in detection and trends—from conventional methods to novel biosensors. Microchim. Acta **186**(6), 329 (2019)
143. Ganguli, A., et al.: Rapid isothermal amplification and portable detection system for SARS-CoV-2. Proc. Natl. Acad. Sci. **117**(37), 22727–22735 (2020)
144. Pascarella, G., et al.: COVID-19 diagnosis and management: a comprehensive review. J. Intern. Med. **288**(2), 192–206 (2020)
145. May, L., Tran, N., Ledeboer, N.A.: Point-of-care COVID-19 testing in the emergency department: current status and future prospects. Expert Rev. Mol. Diagn. **21**(12), 1333–1340 (2021)
146. Takemura, K.: Surface plasmon resonance (SPR)- and localized SPR (LSPR)-based virus sensing systems: optical vibration of nano- and micro-metallic materials for the development of next-generation virus detection technology. Biosensors **11** (2021). https://doi.org/10.3390/bios11080250
147. Das, S.K.: The pathophysiology, diagnosis and treatment of corona virus disease 2019 (COVID-19). Indian J. Clin. Biochem. **35**(4), 385–396 (2020)
148. Boguszewska, K., et al.: Immunoassays in DNA damage and instability detection. **76**(23), 4689–4704 (2019)
149. Adam, H., et al.: Integration of microfluidic channel on electrochemical-based nanobiosensors for monoplex and multiplex analyses: an overview. J. Taiwan Inst. Chem. Eng. **146**, 104814 (2023)
150. Zhao, Y., et al.: Applications of fiber-optic biochemical sensor in microfluidic chips: a review. Biosens. Bioelectron. **166**, 112447 (2020)
151. Xu, X., et al.: Digital microfluidics for biological analysis and applications. Lab Chip **23**(5), 1169–1191 (2023)
152. Antiochia, R.: Paper-based biosensors: frontiers in point-of-care detection of COVID-19 disease. Biosensors **11** (2021). https://doi.org/10.3390/bios11040110
153. Tayyab, M., et al.: Potential microfluidic devices for COVID-19 antibody detection at point-of-care (POC): a review. IEEE Sens. J. **21**(4), 4007–4017 (2021)
154. Busa, L.S., et al.: Advances in microfluidic paper-based analytical devices for food and water analysis. Micromachines **7** (2016). https://doi.org/10.3390/mi7050086

Chapter 6
Advancing Biological Research and Drug Development Through Organ-On-A-Chip Technology

Abstract In this chapter, the groundbreaking concept of organ-on-a-chip technology was introduced and its importance in biomedical research was examined. The use of organ-on-a-chip platforms offered a unique opportunity to mimic complex organ microenvironments and study organ function in a controlled laboratory environment. The fabrication techniques and materials used in organ-on-chip systems were discussed, highlighting the advantages of this technology, including its potential for high-throughput screening, reduced animal testing, and the ability to replicate disease conditions. However, the challenges and limitations associated with organ-on-a-chip models were also acknowledged. The design concepts of specific organs on chips, such as the heart, lung, brain, kidney, liver, intestine, vasculature, and skin, were briefly discussed, explaining their unique functions and contributions to disease modeling and drug development. Additionally, the application of organ-on-a-chip technology in cancer research was examined, with a focus on tumor-on-a-chip models and their materials. Finally, the emerging human-on-a-chip concept, which integrated multiple organ systems into a single device and enabled more comprehensive studies, was presented.

Keywords Biology · Drug development · Organ-on-a-chip · Organ microenvironment · Tumor-on-a-chip · Human-on-a-chip

6.1 Introduction

Microfluidics has attracted extensive considerations as it can propose numerous profits in industrial and biomedical applications [1–4]. Advancements in microfluidic culture technology have led to the development of organ-on-a-chip (organ chip) platforms, which have become valuable tools for studying pathophysiology, tissue development, and human physiology modeling in vitro [5–8]. These microfluidic devices, commonly known as organs-on-chips, provide a distinct culture environment where living cells can be cultured in micrometer-sized chambers, allowing for the replication of physiological functions observed in organs and tissues [9–12]. By

S. M. Naghib and Y. Rezaian, *Frontiers of Nanobiotechnology*, Series in BioEngineering, https://doi.org/10.1007/978-3-031-88419-1_6

replicating the dynamic interactions found in living microenvironments, organs-on-chips, also referred to as OOCs, have emerged as miniature models of organs and tissues, offering insights into human physiology and disease. They provide a relevant context for studying cellular responses within human organ and tissue settings. Organs-on-chips are sophisticated microfluidic cellular cultivation apparatuses that endeavor to emulate the complex multicellular structure, tissue-tissue interfaces, and varied physical microenvironments that are present in living organs [5, 13, 14].

These devices consist of parenchymal and vascular compartments lined with living human cells, offering precise control over parameters such as tissue-organ interactions, tissue boundaries, cell patterning, shear force, and concentration gradients [15]. The goal is to simulate the physiological environment of human organs by creating minimal functional units that mimic tissue- and organ-level functions, rather than attempting to construct complete organs [9]. Organs-on-chips provide valuable insights into the molecular-scale activities occurring in living cells, enabling the investigation of how physical, chemical, cellular, and molecular cues influence human tissue development and disease [1]. Organs-on-chips can exhibit varying levels of complexity, ranging from single-perfused microfluidic chambers containing specific cultured cells to more intricate designs. These designs feature multiple microchannels connected by porous membranes, with each channel lined by different cell types. The primary purpose of these designs is to mimic interfaces between different tissues, such as the blood–brain barrier or the lung alveolar-capillary interface [16].

These systems provide a means to assess specific reactions within organs, particularly the mobilization of circulating immune cells in response to toxins, pharmaceuticals, or other external disruptions. By studying the immune cell response within organs-on-chips, researchers can gain insights into how these cells react to different substances or disturbances and their potential role in combating them [9]. The development of organs-on-chips has been made possible by developments in tissue engineering, semiconductor fabrication, and the procurement of human cells, namely by improvements in human stem cell technologies. Organs-on-chips provide a favorable approach to replicate human pathophysiology in a laboratory setting and overcome the constraints of current cellular and animal models [17].

In this chapter, our objective is to present a comprehensive and captivating overview of the fascinating world of organs-on-chips. We will delve into the intricacies of their design principles, functionalities, and applications in studying tissue development, organ-level functions, and disease modeling. Moreover, we will explore the convergence of microfluidics, tissue engineering, and human stem cell technology that has paved the way for the development of these remarkable bio-mimetic devices. Furthermore, we will examine the potential impact of this ground-breaking technology in bridging the gap between in vitro studies and clinical applications, ultimately advancing our understanding of human biology and drug discovery, and personalized medicine.

6.2 Why Do We Use the Organ on the Chip?

Conventional two-dimensional (2D) cell cultures, which were created nearly a century ago, have been crucial foundations for research in the life sciences [9, 18]. Their technique involves establishing a single layer of cells in a culture flask and adding drugs for assessment into the cell culture solution. Despite their ability to promptly evaluate drug-cell interactions, these 2D systems have shortcomings that hinder their ability to accurately emulate the in vivo surroundings and forecast tissue efficacy and drug performance [15]. Two-dimensional (2D) cell culture systems have limitations in replicating the complex characteristics of living tissues [19–23]. They cannot accurately reproduce the tissue-specific functions, mechanical properties, extracellular matrix (ECM) interactions of many cell types, and the biochemical environment. In addition, these two-dimensional models cannot accurately show the kinetics of real drug release, so the drug doses may be effective in two-dimensional but not in reality in the human body. This model also fails to simulate physiological manifestations, microenvironmental factors present in living tissues, and intra-organ interactions. Consequently, they often require validation using in vivo animal models for further confirmation [9, 18, 24].

In an effort to address the limitations of two-dimensional (2D) cell cultures, researchers have turned their attention to three-dimensional (3D) culture models [25]. Figure 6.1 shows 2D and 3D cell culture. These models, which have been developed for over 50 years, utilize hydrogels composed of synthetic polymers or natural extracellular matrix (ECM) molecules. They provide improved expression of cell function and tissue organization compared to 2D models [26]. By employing techniques such as cell sheets, spheroids, and cell aggregations, 3D models better capture the chemical and spatial complexity found in living tissues [2, 4]. However, there are also certain drawbacks associated with 3D models. For example, organoids, a type of 3D model, exhibit significant variability in shape and size, making it challenging to maintain cells in consistent positions for extended periods of analysis. This variability can also hinder the functional analysis of cells within the organoids and impede the sampling of luminal contents for analysis. Many 3D systems lack tissue-tissue interfaces and multiscale architecture, which are vital for the proper functioning of organs. Additionally, cells within 3D models frequently do not encounter typical mechanical stimuli, such as compression, fluid shear stress, and tension, that play crucial roles in organ development and function [26].

While there have been notable advancements in 3D culture models, they still fall short of completely replicating the crucial features necessary for the optimal functioning of living organs. These features include tissue-tissue interfaces, the presence of spatial and temporal gradients of oxygen and chemicals, and a mechanically active microenvironment [25]. To overcome these constraints and offer a more accurate representation of the human body, the organ-on-a-chip technique has surfaced. Organ-on-a-chip setups merge the benefits of three-dimensional culture models with more advanced capabilities that simulate the intricacies of real organs.

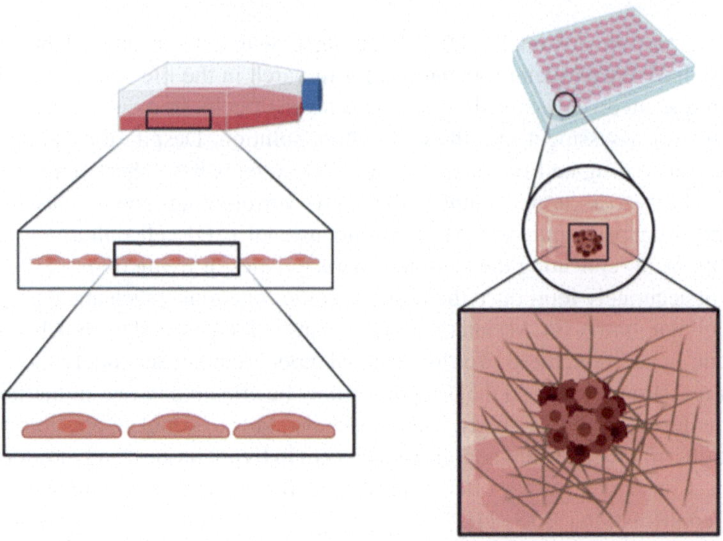

Fig. 6.1 Cell behavior in 2D conventional cell culture vs 3D cell culture: **a**. 2D-cultured cells often have a flat shape that does not reflect the genuine physiological cell morphology. **b** Cell behavior in three-dimensional cell cultures. Cells grown in three-dimensional systems display more realistic appearance and behavior because they develop in a microenvironment similar to that found in the body. Reprinted with permission from Ref. [26]. Copyright 2021, MDPI

These microfluidic contraptions strive to duplicate the multi-cellular design, tissue-tissue connections, and physiological microenvironments that exist in actual organs [15].

Organ-on-a-chip platforms present a multitude of advantages compared to animal models and traditional cell cultures. By incorporating human cells, these platforms offer a more precise portrayal of human biology, thereby facilitating the investigation of human-specific responses and illnesses [25]. Furthermore, organ-on-a-chip platforms serve as a valuable tool in addressing the limitations of 2D and 3D cell cultures, as they integrate the intricate spatial and chemical characteristics of living tissues. The primary objective is to replicate tissue-tissue interfaces and simulate mechanical cues that are fundamental in the proper development and operation of organs [13]. Sophisticated microfluidic techniques are employed in these systems, enabling the investigation of intercellular interactions between cultured cells and circulating blood or immune cells [9]. Although animal models continue to be the preferred choice for in vivo studies, they do have certain limitations, including phylogenetic differences and a lack of accuracy when it comes to predicting human responses [25]. Additionally, using animals for preclinical testing is associated with ethical concerns, high expenses, and time-consuming procedures. Organ-on-a-chip technology offers an alternative approach that can work in tandem with animal models, providing a more human-specific and predictive platform for toxicity testing and drug development [18].

In conclusion, the technology of organ-on-a-chip tackles the constraints of animal models and conventional cell culture systems by offering a more physiologically significant platform for investigating human biology, simulating diseases, promoting drug discovery, and nurturing personalized medicine methodologies. Through emulation of the intricacies of living organs, organ-on-a-chip frameworks possess the capacity to transform the field of biomedical research and augment human health outcomes.

6.3 Engineering the Future: Fabricating Organs on a Chip

The creation of organ-on-a-chip models encompasses a range of methods, each with its unique benefits and obstacles. One of the original approaches, which drew inspiration from photolithographic etching methods utilized for computer microchips, involves a modified version of this process to manipulate surface features on a scale that cells can perceive and react to. Additionally, soft lithography, a technique that duplicates patterns found on silicon chips using biocompatible materials, has also been utilized [27]. This technique entails pouring a liquid polymer like PDMS onto an etched silicon substrate to generate a rubber-like substance that serves as the foundation of the chip. Initially, soft lithography was employed to produce small, specified regions that were coated with adhesive molecules or extracellular matrix (ECM) proteins on silicon chips and other substrates [28]. These particular areas acted as precise attachment points for cells, directing their placement and function within the microfluidic device or tissue culture system. Subsequently, this method was modified by flipping over the PDMS mold and sealing it to a smooth substrate, generating small, hollow chambers or microfluidic channels for fluid perfusion. These microfluidic channels operate as miniaturized perfusion bioreactors where cells stick to ECM-coated surfaces and are consistently exposed to the culture medium [9].

Although these techniques have demonstrated their efficacy, they are not without limitations. The fabrication process necessitates the employment of multiple lithographic steps, specialized equipment, and labor-intensive procedures, ultimately rendering it both costly and time-consuming. Furthermore, these methods primarily concentrate on producing the microfluidic chip itself, while additional components such as microtissues, stimulus-loading elements, and results-readout sensors necessitate further processing [29]. The difficulties associated with these approaches become significant when implemented in biomedical research and cell biology studies within microfluidic systems. The extended preparation process can adversely affect cell viability and functionality. Additionally, issues such as inadequate support for cell attachment and precision in cell patterning hinder the effectiveness of organ-on-chip systems. To overcome these obstacles, 3D fabrication techniques have been explored, utilizing microfabrication methods to construct intricate tissue architectures [30].

Newly, the technology of 3D bioprinting has emerged as the most advanced method for the fabrication of microfluidic devices (Fig. 6.2) [31]. This technology has been applied to the development of Organ-on-a-Chip (OOC) systems due to significant methods, rapid generation of microfluidic channels with user-friendly equipment, high efficiency, and processing versatility, that have been developed. These methods utilize cells, bioactive molecules, and various natural bio-inks to construct 3D tissue models in vitro. Bioprinting can be executed through various modes, such as laser-based (LBB), droplet-based (DBB), and extrusion-based (EBB) bioprinting, either in a scaffold-free or scaffold-based fashion [32]. In the future, the development of organ-on-a-chip (OOC) systems with human physiological and anatomical features will greatly benefit from the rapid and continuous model generation, testing, and redesign capabilities offered by 3D bioprinting technology. This advanced technology enables the creation of OOCs that closely mimic the interactions between organs in the human body. One notable advantage of OOC approaches is their ability to significantly reduce the resources required for evaluation in terms of cost and time [33].

Various organs, including the kidneys, heart, intestines, blood vessels, liver, lungs, and even tumor microenvironments, have been extensively studied and successfully replicated using these techniques. The replication of these organs within OOC systems allows for more accurate and reliable testing, opening up new possibilities for biomedical research and drug development (Fig. 6.3) [25, 34]. Bioprinting enables the accurate placement of multiple materials with exceptional spatial resolution. Furthermore, mechanical stimuli are integrated through the printing of actuators on chips or by employing smart materials in 4D bioprinting, effectively applying external stimuli to the microtissues on the organ-on-a-chip [29]. The application of these fabrication techniques has led to remarkable progress in creating in vitro tissue models for both biomedical research and drug development. However, there are still several obstacles that must be overcome, including the prolonged preparation process that may impact cell viability and functionality, together with concerns regarding cell attachment and pattern precision [32].

In summary, the advancement of organ-on-a-chip models encompasses diverse construction techniques, all of which possess their own benefits and challenges. These techniques aid in the progression of in vitro tissue models, yet additional enhancements are necessary to surmount current limitations and heighten the efficacy of organ-on-a-chip systems.

6.4 Materials in Organ-On-A-Chip Platforms

In this division, we will categorize materials used in organ-on-a-chip (OOC) platforms into two distinct categories: the materials employed in chip construction and the biomaterials utilized for tissue fabrication.

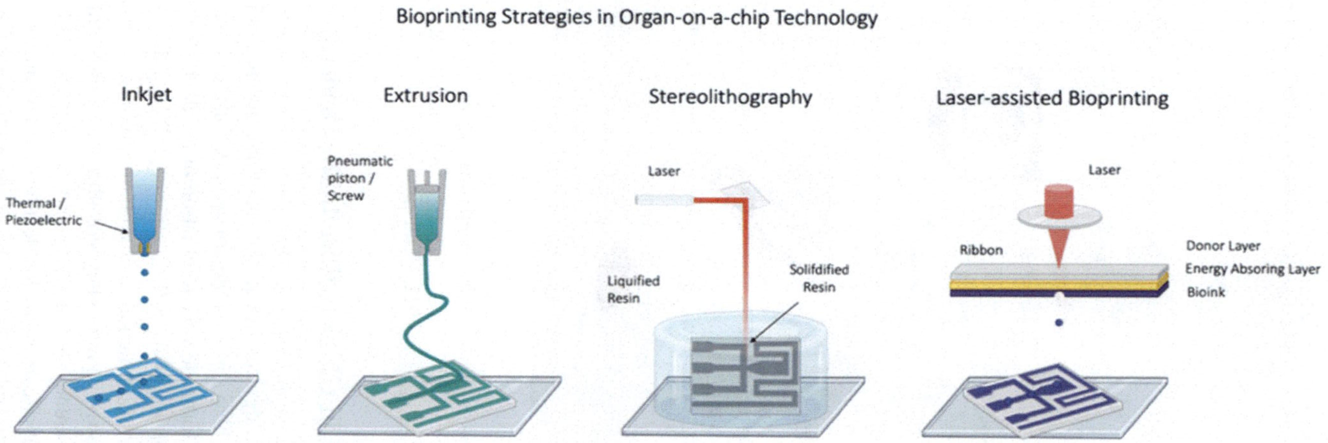

Fig. 6.2 Bioprinting technologies shown schematically include inkjet, extrusion (FDM), stereolithography (SLA), and laser-assisted bioprinting. Reprinted with permission from Ref. [31]. Copyright 2022, Frontiers

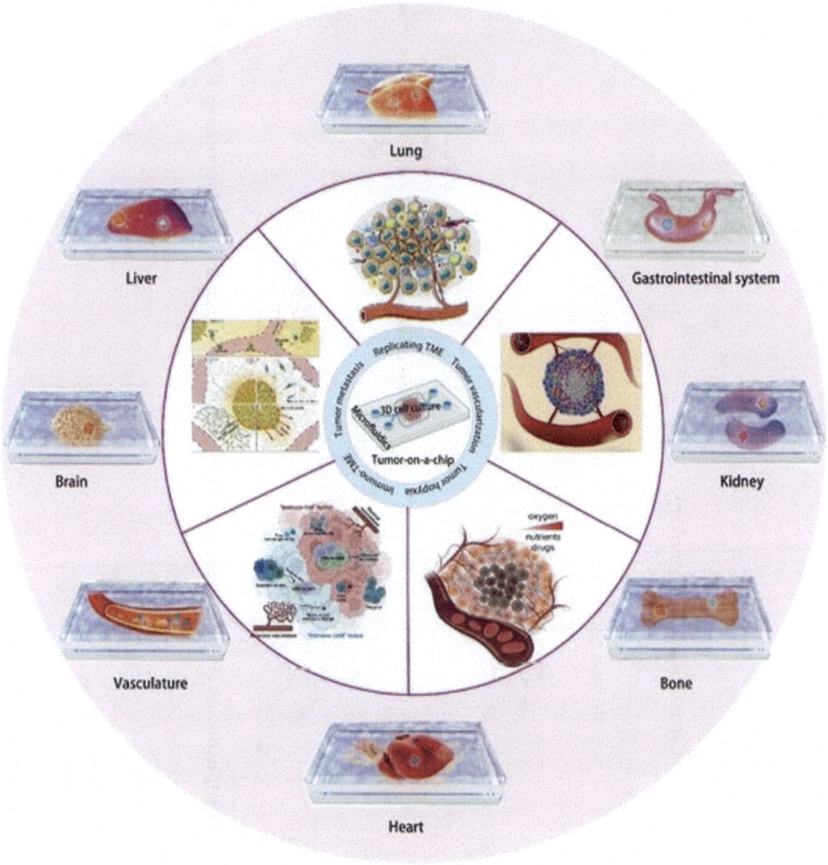

Fig. 6.3 Diagram showing the applicability of OOC platforms for organ reconstruction on microfluidics for cancer research, as well as their ability to imitate the TME and functions in vitro. Reprinted with permission from Ref. [34]. Copyright 2022, MDPI

6.5 Materials Used in Chips (Microfluidic Devices)

The very essence of OOC devices lies in the materials that not only facilitate the attachment of tissues but also serve as the framework for the device channels and features. These materials form the very foundation of the device's construction and are instrumental in supporting its overall functioning. Glass: Glass substrates offer exceptional optical properties, which allow for compatibility with microscopy techniques and high-resolution imaging. They are utilized in microfluidic chips that necessitate optical transparency for analysis and observation. Polymers: Various types of polymers find application in the design of microfluidic devices. These include poly (methyl methacrylate) (PMMA), cyclic olefin copolymer (COC), and polystyrene (PS). Each of these polymers has unique properties that cater to specific requirements. For instance, PS and PMMA offer biocompatibility and transparency, respectively, while COC provides excellent chemical resistance. Silicon: In certain microfluidic applications, where integration with electronic components or sensors is desired, Silicon reigns supreme. The fabrication of Silicon-based microfluidic chips is possible through standard Silicon processing techniques, thereby enabling the seamless integration of electronic elements [35, 36].

Polydimethylsiloxane (PDMS): Polydimethylsiloxane (PDMS) is a synthetic polymeric elastomer based on silicone and is widely used for chip fabrication. Due to its flexibility, ease of fabrication, biocompatibility, and transparency, PDMS is a commonly used material in microfluidic devices. The substrate and microchannels of the chip are formed by PDMS, enabling the flow of fluids and supporting cell culture. The creation of PDMS chips often entails the utilization of a methodology referred to as replica molding. This involves the production of a microfabricated design via soft lithography, and then generating replicas by pouring PDMS onto a master mold. The PDMS undergoes crosslinking, causing the chip device to solidify. Initially, the PDMS solution is poured over the features of the master mold. After drying, the solidified PDMS layer is peeled away, revealing the mold. For further processing, the mold can be enhanced by covering it with either a glass slide or an additional PDMS block. PDMS prototypes find application both as on-chip devices and as molds for secondary materials. While PDMS offers a convenient fabrication process and adjustable elasticity, it does have certain limitations. The hydrophobic nature of PDMS requires surface modification to facilitate cell attachment. Although surface treatments, like oxygen plasma treatment, can enhance attachment, their effectiveness is not always complete. Additionally, PDMS devices have high permeability to small hydrophobic molecules, which can affect solution concentrations and lead to water evaporation. Moreover, the non-degradable nature of PDMS limits its applicability in specific contexts [37, 38].

6.6 Biomaterials Used in Tissue Fabrication

Biomaterials that are utilized in OOC platforms are frequently endowed with porous structures and biodegradable properties. Biomaterials can be obtained from various sources and exhibit specific interactions with cells and other biological components within the body. These sources include synthetic materials, such as polylactide and polyglycolide, natural materials, such as collagen and fibrin, or hybrid natural-synthetic materials. Collagen: Collagen is a remarkable protein that naturally exists in the extracellular matrix (ECM) of tissues. Its ability to promote cell adhesion, migration, biodegradability, and biocompatibility, differentiation makes it a highly desirable biomaterial. Collagen, available in various forms, can be used as a scaffold or incorporated into hydrogels for tissue engineering purposes [39].

Alginate: Alginate, a naturally occurring polysaccharide obtained from seaweed, has the unique ability to form hydrogels upon interaction with divalent cations such as calcium. This property enables the encapsulation and delivery of cells, providing a suitable 3D environment for cell growth and tissue formation. The formation of alginate hydrogels has significant implications in the field of regenerative medicine due to their biocompatibility and ability to mimic the extracellular matrix. Gelatin: Gelatin, which is obtained from collagen, possesses comparable characteristics such as biocompatibility and facilitation of cell adhesion. This material serves as an ideal scaffold and can be seamlessly integrated into hydrogels for diverse tissue engineering applications. Synthetic Polymers: Various synthetic polymers, including poly(caprolactone) (PCL), poly (lactic-co-glycolic acid) (PLGA), and polyethylene glycol (PEG), have found prominent use in the field of tissue engineering. These polymers possess the ability to exhibit surface characteristics, degradation rates, and tunable mechanical properties, thereby enabling them to cater to specific tissue engineering requirements with remarkable precision [40].

Decellularized Extracellular Matrix: Decellularized Extracellular Matrix (DECM) refers to the complex environment obtained by removing cellular components from tissues, resulting in the preservation of the native ECM structure. This structure contains bioactive molecules, growth factors, and structural cues that facilitate cell attachment, tissue regeneration, and migration. The materials used in the production of chips and biomaterials for tissue fabrication offer numerous opportunities for creating microfluidic devices and engineering functional tissues. The selection of materials depends on various factors, such as biocompatibility, desired characteristics, intended applications, and specific requirements of the tissue or organ being investigated or constructed [41].

6.7 Materials for Bioprinting

The use of bio-ink, a combination of cells and printable biomaterial, requires specific essential properties such as biodegradability, high biocompatibility, bio-mimicry, structural integrity, and sufficient mechanical, and appropriate rheological properties to ensure effectiveness in the printing process [42–44]. Various hydrogels have been employed as sources of bio-ink, including synthetic and natural polymers, each with its own advantages and disadvantages. These materials possess unique characteristics, and their continuous development has improved their relevance in tissue engineering [20, 23, 45]. Natural bio-inks, especially those originating from animal sources like hyaluronic acid, fibrin, collagen, gelatin, and their mixtures, possess the ability to enhance biological features by providing physical and biochemical stimuli to guide cellular behaviors. However, these bio-inks do possess certain limits concerning variability across batches, structural durability, and the possibility of disease transmission. The cellular interactions with the plentiful matrix, such as the extracellular matrix, might result in the formation of fresh tissue via biochemical processes. The exchange of biochemical signals between cells and their surrounding matrix, including the extracellular matrix, can result in the generation of fresh tissue. This exchange encompasses a communication process that prompts underlying signaling mechanisms. In essence, the cells and matrix collaborate by exchanging signals and cues to initiate and promote the formation of new tissue [8, 46, 47].

On the other hand, Alginate rapidly undergoes gelation when in contact with a mechanically sturdy substrate, providing stability to the printed structures. However, it is not highly suitable for cell encapsulation due to its relatively low affinity for cells. The techniques used for cell encapsulation often neglect the usage of synthetic polymers such as PLGA and PCL due to their high melting temperatures. Nevertheless, these polymers possess the advantage of generating mechanically robust structures. However, they suffer from the inherent drawback of having a reduced number of active binding sites, leading to a decrease in cell adhesion and limited bio-functionality. To combat these challenges, researchers have explored the potential of chemical conjugation involving both synthetic and natural polymers. For instance, PEGylated gelatin methacrylate (PEGgelMA) or the amalgamation of gelatin-methacrylate (GelMA) can significantly enhance the strength of hydrogels through post-polymerization techniques. Additionally, the combined effect of these polymer blends displays encouraging results in improving cellular responses. The high bio-mimicry of tissue-specific microenvironments in engineered tissues makes decellularized extracellular matrix (DECM) a highly promising bio-ink with significant potential. The complexity and composition of proteins, including glycoproteins and proteoglycans, within the extracellular matrix (ECM), are vital for successfully mimicking intricate tissue functions that are difficult to achieve through conventional methods. Notably, the abundant collagenous proteins found in DECM possess a significant characteristic a temperature-responsive crosslinking process which proves to be crucial during the printing process of DECM bio-ink. This unique property allows for precise control and manipulation, facilitating the creation of tissue constructs with enhanced structural integrity. Recent research by Pati and colleagues [48] has led to the development of printable tissue-specific DECM bio-ink, with subsequent investigations on the liver, skeletal muscle, fat, and heart demonstrating superior tissue maturation compared to traditional ECM bio-inks [49–51].

6.8 Advantages of Organ on Chip Technology

Organ-on-Chip technology offers numerous advantages that differentiate it from conventional approaches to studying human cells and tissues. One notable advantage lies in the cost-effectiveness of manufacturing Organ-on-Chip devices, enabling even researchers and institutions with limited resources to utilize this technology. Another significant benefit is the precise testing capability of a broad range of drug concentrations, which grants researchers more accurate and comprehensive insights into the effects of different dosage levels. Organ-on-Chip devices truly shine when compared to the conventional practice of using Petri dishes. They demonstrate exceptional capabilities in replicating the intricate micro-environment of human cells, ensuring a remarkably accurate portrayal of essential factors like pH, temperature, and oxygen levels. As a result, the experimental outcomes derived from Organ-on-Chip technology offer a heightened level of realism, empowering researchers with a deeper understanding of cellular behavior and responses [52, 53].

6.9 Disadvantages of Organ on Chip Technology

Adsorption on Inner Linings: The analysis quality in Organ-on-Chip devices may be impacted by the adsorption of products on the inner linings. This phenomenon of adsorption may cause certain substances to be lost or changed, which can jeopardize the precision and dependability of the analysis outcomes. It is imperative to tackle this problem to guarantee the validity of experimental data. Limited Mixing Potential: One disadvantage of Organ-on-Chip technology relates to its limited mixing potential. The flow dynamics within these devices often exhibit a small Reynolds number, leading to predominantly laminar flow patterns. This laminar flow condition poses a challenge to efficient fluid mixing, potentially impacting the uniform distribution of substances and impeding certain experimental processes. Researchers are actively working to find effective solutions that promote enhanced mixing capabilities within Organ-on-Chip devices. By addressing this limitation, the technology can further improve its performance and broaden its applications in the study of tissues and human cells. It is important to acknowledge that the pursuit of mitigating these shortcomings is an active subject of research and development in the realm of Organ-on-Chip technology. The objective of researchers is to surpass these constraints in order to amplify the efficacy and versatility of this pioneering method of investigating human cells and tissues [52, 53].

6.10 Approaches to Design Organs-On-Chips

The intricate growth of cells relies on a delicate interplay between various complex internal and external environmental factors. When developing in vitro physiological models, it becomes crucial to consider the actuality of external environmental parameters. This is where the integration of microfluidic technology, micromachining, and cell biology proves invaluable. Through this combined approach, organs-on-chips allow for precise regulation of environmental parameters. This regulation encompasses fluid biochemical concentration gradients, mechanical stress, shear stress, and a range of other chemical and physical stimuli. The cells contained within the chip are capable of responding to these stimuli, thereby promoting self-organization and exhibiting more genuine physiological activities. These unique benefits of organ-on-chip technology underscore its pivotal role in establishing in vitro physiological models, enabling researchers to gain deeper insights into cellular behavior and functions [10, 17].

(1) The act of generating fluid shear stress is a perpetual occurrence within the human body, yet it is non-existent in the conventional static culture. With the aid of microfluidic technology, dynamic perfusion in cell culture can be achieved through the use of micro pumps, which ultimately promotes consistent nutrient delivery and prompt waste removal. Moreover, this dynamic cell environment is far more comparable to that of the human body. Additionally, the fluid shear stress present in perfusion culture is paramount in facilitating certain physiological functions such as renal reabsorption.
(2) The creation of dynamic mechanical stress is an intriguing process. The pressure within the human body is closely tied to a variety of vital life functions, such as skeletal stress, pulmonary pressure, and blood pressure. Consistent pressure plays a significant role in the body's physiological processes, including cell differentiation, tissue formation, and even the formation of tumors. By leveraging microfluidic technology, it is possible to generate periodic mechanical stress through the use of a flexible, porous membrane on which cells can be cultured. External forces can cause the membrane to undergo mechanical stretching, which can simulate certain physiological functions such as heart contractions, lung breathing, and intestinal peristalsis.
(3) The microscale fluid flow is characterized by laminar movement, which is advantageous in creating various types of concentration gradients within the channel. These biochemical signals, driven by different concentration gradients, play a crucial role in numerous physiological processes, including immune response, cancer metastasis, migration, and cell differentiation. Microfluidics technology facilitates the creation of stable biochemical concentration gradients in a 3D manner through the manipulation of channel size, flow rate, and the implementation of distinctive channel configurations like micropumps or microvalves. Moreover, this advancement enables the simulation of intricate physiological processes occurring within the human body. Furthermore, the generation of multi-channel concentration gradients has greatly expedited the process of conducting high-throughput drug screening on a microchip.
(4) The orchestration of cell patterning is a fundamental aspect of the human body's complex unity, where a variety of cells are intricately arranged to form functional units through dynamic interaction. Microfluidic technology has proven to be a distinguished tool for controlling cell patterning. The use of the micropillar array, laminar flow, electrochemical technique, surface modification, and template technique have all played a role in cell patterning on the chip. This progress is extremely beneficial in creating a physiological model in the laboratory with a complex geometric structure, while also offering an outstanding opportunity to study interactions between cells and their surrounding matrix as well as interactions between different cells [54].

6.11 Single Organs-On-A-Chip

The concept of Single Organs-on-a-Chip is an advanced and innovative approach that aims to replicate the physiology of specific organ systems within microfluidic devices. The design principles focus on utilizing the simplest functional unit of each organ system to create an environment that closely mimics the targeted organ. By incorporating biophysical stimuli and advanced microfabrication techniques, Single Organs-on-a-Chip devices can recreate the complexity and functionality of the organ in question. These miniature organ models integrate multiple cell types, vascular networks, and relevant biochemical cues to emulate the intricate interactions within the human body. The potential applications for this technology are immense, including disease modeling, drug testing, and personalized medicine. It is revolutionizing biomedical research and facilitating a deeper understanding of organ physiology while offering a platform for more accurate evaluation of therapeutic interventions [55].

6.12 Lung-On-A-Chip

The lung is a crucial organ responsible for the exchange of oxygen and carbon dioxide in the bloodstream. Investigating blood flow, cellular interactions, and gas flow within the respiratory tract is essential for drug delivery testing and physiological research. At the most minuscule scale, the alveolar-capillary membrane functions as the quintessential constituent of the lungs, primarily composed of microvascular endothelial cells and alveolar epithelial cells. This membrane enables efficient gas exchange, as carbon dioxide diffuses from the bloodstream into the alveoli while oxygen diffuses from the alveoli into the arterial blood vessels. Furthermore, it acts as an impenetrable barrier, inhibiting the ingress of pathogens and detrimental particles into the pulmonary circulatory system. The interface comprises specialized epithelial cells that secrete mucus-containing substances with antibacterial properties and signaling chemicals that reduce inflammation. These chemicals enhance the migration of immune cells to contaminated areas [56]. Microfluidics technology facilitates the development of artificial lung models and the investigation of lung diseases through precise manipulation of fluid flow and gas exchange. Ongoing research is primarily concerned with studying the blood-blood barrier (BBB), examining the impact of shear force on pathological mechanisms, and regulating airway mechanical pressure [18].

Huh and colleagues initially introduced the pioneering lung-on-a-chip architecture [15, 57]. Their work involved the development of a biomimetic microsystem that emulates the crucial functional interface between alveoli and capillaries in the human lung, which represents the primary functional unit of the respiratory organ [58]. A lung-on-a-chip system consists of a membrane that divides one main channel and two side channels. Membrane and lateral channels form the main components of this system [15]. The membrane's upper regions are occupied by alveolar epithelial cells, and the lower regions contain pulmonary microvascular endothelial cells. This configuration effectively replicates the alveolar-capillary barrier, enabling the investigation of the impact of mechanical forces and cell–cell interactions. The device mimics the contraction and expansion of the alveoli during breathing by using adjacent microchannels to apply controlled mechanical stretching [18]. The innovative lung-on-a-chip technology introduced by Huh and colleagues has demonstrated its effectiveness as a valuable means of exploring diverse aspects of lung biology. This method facilitates the observation and characterization of inflammatory processes, which allows researchers to scrutinize the reaction to bacteria and the emergence of pulmonary edema. As a matter of fact, this system has been utilized successfully to emulate drug-induced pulmonary edema, providing insights into its fundamental mechanisms and assessing conceivable remedies [56].

In addition, the use of lung-on-a-chip systems has enabled the examination of how lung tumor cells react to alterations in their microenvironment. This has led to a greater comprehension of lung cancer and possible treatment methods. Through the application of computational techniques such as the finite element method and computational fluid dynamics, researchers can optimize the device's shape and flow rates to enhance the precision and physiological significance of these microsystems. Lung-on-a-chip technology is incredibly promising for expanding our understanding of lung biology, studying diseases, and assessing how drugs work. These devices create a controlled environment that closely mimics the conditions in the human lung, allowing for more precise and effective drug development, personalized medical treatments, and reduced reliance on animal testing. However, as this field progresses, it is crucial to address regulatory and ethical concerns to ensure that the results obtained from lung-on-a-chip models are reliable and can be reproduced accurately.

6.13 Gut-On-A-Chip

The crucial function of the human gut in digestion, nutrient absorption, and hosting an intricate microbial ecosystem is undeniable [59]. In vitro models play a crucial role in the study of pharmacokinetics for the human intestine. However, traditional models fail to replicate the physiological microenvironment of the gut, including cyclic peristaltic motion. To address this issue, microfluidic platforms have been introduced as a more powerful in vitro model for the intestine, offering a more realistic environment for drug testing. The gut-on-a-chip aims to emulate the essential factors of the gut microenvironment, which include the intestinal epithelial barrier, the presence of intestinal cells, and the interaction with gut microbiota [59]. Gut-on-a-chip (OOC) models have been developed by adopting the lung-on-a-chip concept, which mimics the villi-rich surface area of the gut and incorporates microbial flora. This innovative technology involves a transparent microfluidic device that encompasses multiple channels, meticulously lined with human intestinal cells, such as endothelial cells and epithelial cells, which form a thin layer to simulate the intestinal barrier [60].

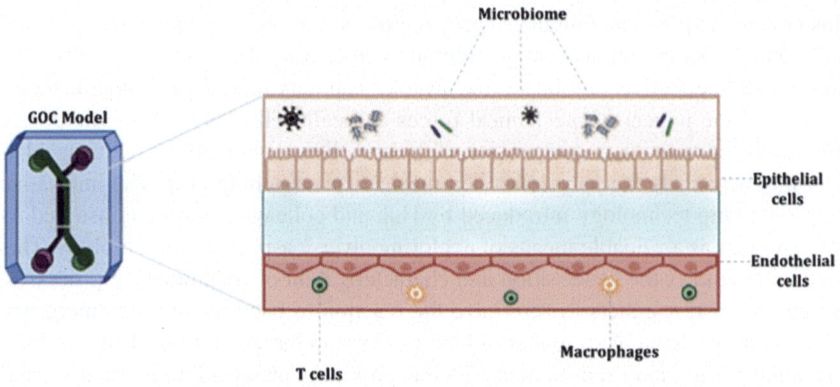

Fig. 6.4 Diagrammatic illustration of the GOC model that resembles the gut microenvironment. Reprinted with permission from Ref. [61]. Copyright 2023, MDPI

Additionally, membrane-based OOCs with 3D structures have been utilized to simulate villi formation and enable access to both basolateral sides of the gut epithelium and apical (Fig. 6.4) [61].

The integration of stirring pumps into gut-on-a-chip systems has been undertaken by researchers to regulate fluid flow and foster a dynamic environment. Kim et al. [62, 63], for example, designed a porous membrane-based dual-lumen microfluidic device, where Caco-2 cells were cultured for a prolonged period, and Rhodamine 123 was employed to investigate cell permeability. In a recent study, the peristaltic motion and fluidic dynamics of the intestine were successfully imitated, using Caco-2 cells in flexible microfluidic channels. This method led to the emergence of undulating epithelium columns comprising polarized Caco-2 cells and multiple distinct types of intestinal cells, including enteroendocrine cells, Paneth cells, and Goblet cells [58]. The successful absorption of oral drugs relies on their traversal through the small intestine into the bloodstream. Key to this process is the villi, which play a crucial role in absorption, and it is essential to maintain their morphology on the chip. Researchers, such as Imura and his colleagues, have developed chips that aim to simulate the intestinal system. These chips consist of a glass slide permeable membrane and a PDMS sheet containing channels where Caco-2 cells can be cultured. Sung and his colleagues [64] have introduced a novel 3D hydrogel structure that mimics the intricate structure of human intestinal villi. Additionally, Kim and his colleagues have developed bionic devices, which recreate the microenvironment of the intestine using cyclic strains and shear force. Caco-2 cells have shown remarkable growth and have been able to maintain the microbial flora present in the human intestine within these chips. This complex system, with its intricate structure and physiological characteristics, has provided a valuable platform for the investigation of the intestinal microbiome and theevaluation of drugs, inflammatory cells, and mechanical deformation associated with peristalsis in the context of intestinal diseases. The application of this apparatus has granted researchers the opportunity to delve into the origins of intestinal diseases, identify potential targets for therapeutic interventions, and evaluate the effectiveness of medications. These discoveries underscore the promising potential of intestine-chip models in the realm of personalized medicine studies that focus on the intricacies of intestinal cells. Intestinal cells can be cultivated either in isolation or in conjunction with endothelial cells, which may include HUVECs [18].

The researchers conducted a comprehensive examination of the inflammation of the intestine resulting from excessive bacterial growth. This endeavor afforded them the opportunity to thoroughly investigate the pathophysiology of this ailment over the course of multiple weeks, rendering it an exceptional model for diverse medical applications. Exposure to ionizing gamma radiation during therapy can give sepsis, rise to intestinal hemorrhage, and ultimately fatality. In order to assess the efficacy of medications, scientists frequently subject cells from murine models to radiation, causing harm and quantifying reactive oxygen species. Nonetheless, the utilization of a gut-on-a-chip approach for simulating gamma radiation-induced injury presents a feasible alternative to conventional in vitro testing methodologies. This helps to maintain the continuity of cellular junctions and yield comparable results. The gut-on-a-chip model has found yet another remarkable application in replicating the CVB1 virus infection. This virus is notorious for causing myocarditis, infecting the liver and pancreas, and leading to severe complications, particularly in neonatal cases. As there is an urgent need to explore this field using human cell lines, the human chip serves as a dynamic system that closely mimics the microenvironment and facilitates successful testing of a polarized infection. Villenave and his colleagues [65] research showcases how this model can be employed to test other enteroviruses and emphasizes the significance of utilizing complex systems over static cell cultures to establish more precise outcomes [58].

6.14 Liver-On-A-Chip

The liver, the largest internal organ in the human body, plays a crucial role in various metabolic functions such as detoxification, glycogen storage regulation, red blood cell decomposition, plasma protein synthesis, and hormone production. It is primarily composed of hepatocytes, accounting for 70% to 85% of its mass, and these cells are responsible for its main functional activities [66, 67]. Although the liver possesses remarkable regenerative capabilities, it can be impaired by drug reactions (e.g., ibuprofen and aristolochene) and chronic diseases (e.g., alcoholic hepatitis and viral), which hinder its normal physiological functions. While the liver has been studied using different models like 2D primary hepatocytes, liver-on-a-chip, hepatocyte organoids, and hepatocyte spheroids, the liver-on-a-chip has gained prominence in drug development due to its ability to replicate the liver's microenvironment and offer reliable assay robustness, including reproducibility [58, 66]. Developing a microfabricated liver model presents significant challenges but has attracted considerable research interest [24].

Toxicity in vivo is a major factor contributing to the failure of approximately 90% of drugs during Phase I clinical trials, which leads to substantial costs. In vitro models possess the potential to surpass animal models utilized in preclinical trials, in terms of predicting in vivo toxicity. Such models can significantly curtail the failure rate of drugs in the drug discovery process, thereby minimizing late-stage setbacks. These models also address the constraints related to species-specific variations between animal models and human clinical trials. Hence, developing hepatic culture methods that accurately replicate in vivo circumstances is essential, enabling precise assessment of drug metabolism and toxicity. To overcome the limitations of traditional static culture techniques, researchers have turned to the integration of microfluidics and tissue engineering, which has led to the development of a dynamic and suitable environment for in vitro liver cultures. This approach, known as liver-on-a-chip, aims to address the challenges of nutrient supply and maintaining a physiologically appropriate ratio of medium to cells, especially when cells are grown in 2D monolayers. Figure 6.5 [68] illustrates the design and operation of the tri-vascular liver-on-a-chip (TVLOC), which is one example of such a system. By utilizing microfluidic chips, liver-on-a-chip models replicate the liver's microenvironment and offer reliable assay robustness, including reproducibility. This advancement holds great promise for drug development by providing a platform that accurately replicates in vivo circumstances, enabling precise assessment of drug metabolism and toxicity. One crucial aspect in designing an effective culture system is the selection of hepatic cell sources [56]. In a significant study by Kane et al. [69] they demonstrated the pioneering use of microfluidic pores to co-culture rat liver cells and 3T3-J2 fibroblasts, simulating an airway interface. The rat hepatocytes cultured within the chip exhibited continuous and stable synthesis of albumin, along with maintained metabolic activity [18].

Lee and his colleagues devised a microfluidic chip that faithfully reproduced the intricate structure of endothelial cells, enabling the cultivation of primary hepatocytes [70]. By implementing a design that allowed the culture media to flow outside the permeable endothelial gap, hepatocytes were successfully organized into cord-like formations, effectively isolating them from the external sinusoidal region. This innovative approach facilitated efficient substance exchange within the system. Similarly, a researcher employed electrophoresis to create radial electric field gradients, enabling precise cell patterning on circular chips composed of polydimethylsiloxane (PDMS). These pioneering techniques aimed to emulate the complex structure of hepatic lobules. A research team introduced a chip architecture with two layers that utilized a porous membrane made of polyethylene terephthalate (PET) to separate the channels. This particular design allowed for a continuous flow of fibronectin-coated rat primary hepatocytes and collagen into the lower channel via the upper chamber. To further enhance the physiological relevance, scientists have explored techniques to culture hepatocytes in a three-dimensional (3D) manner using microfluidic chips. A research team created a platform that mimicked the natural environment for the perfusion of hepatic spheroids, while a researcher studied the interaction between hepatocytes and other types of cells. Various assays have been devised to evaluate liver cell drug toxicity in a high-throughput manner. A researcher has introduced microfluidic electrochemical chip Immunosensors that possess the ability to detect

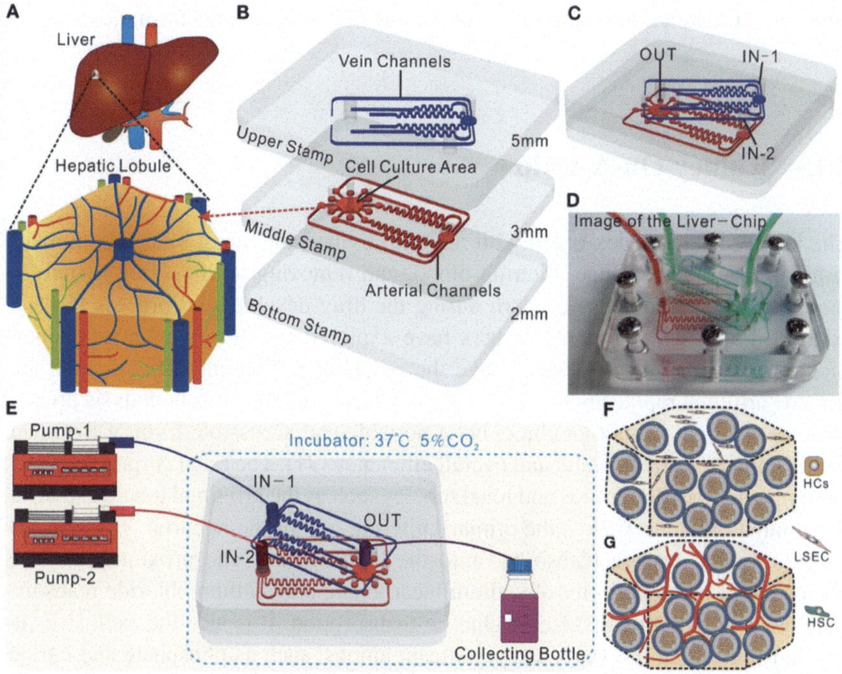

Fig. 6.5 The tri-vascular liver-on-a-chip, or TVLOC, is designed and operated as follows: **a** a schematic of the in vivo hepatic lobule; **b** a schematic of the TVLOC's components; **c** a schematic of the TVLOC assembled; **d** a schematic of the TVLOC's physical picture; **e** a schematic of the TVLOC operation system; **f** a schematic of the cultivation process in the culture area early on; and **g** a schematic of the formation of vascularized liver tissue in the culture area. Reprinted with permission from Ref. [68]. Copyright 2022, frontiers

biomarkers associated with hepatotoxicity. A researcher has developed assays to assess drug-induced skin sensitization by examining metabolite production and the activation of antigen-presenting cells (APCs). This particular system has considerable potential as a drug screening platform for the identification of compounds that could potentially induce systemic skin reactions. In an endeavor to forge more lifelike representations of diseases for the purpose of anti-cancer pharmacological research, a researcher ingeniously merged decellularized liver matrixes (DLM) with gelatin methacryloyl (GelMA) to fabricate a biomimetic liver tumor. This innovative approach effectively emulated the intricate three-dimensional tumor microenvironment (TME), providing an enhanced platform for scrutinizing cancer and evaluating possible therapies. Furthermore, these systems have been employed to explore diverse states of disease and injury. A researcher utilized their system to scrutinize the replication of the hepatitis B virus, whilst another researcher devised a model to investigate the consequences of alcohol-induced injury. Additional advancements can be made by characterizing the cultured cytoplasm through epigenomic analysis, genomics, proteomics, and metabolomics. These techniques hold promise for enhancing the functional outcomes and gaining valuable insights from these studies [18].

6.15 Kidney-On-A-Chip

The kidney is a complex organ with multiple tissues, playing a vital role in maintaining electrolyte balance, filtering blood, and removing toxins. Drug toxicity to the kidney is a significant concern during the drug development process [59]. To address these challenges, researchers have explored the development of kidney-on-a-chip devices. Such devices have the potential to revolutionize research and provide artificial replacements for lost kidney function. The current dialysis process requires frequent visits to a clinic, but a portable and accessible form of treatment would enhance patient health and overall efficiency [71]. The renal corpuscle, which encompasses the glomerulus and renal tubules such as the proximal tubule and distal convoluted tubule, serves as the primary filtering unit of the nephron. Through this process, molecules are reabsorbed into the bloodstream. The proximal tubule is responsible for the majority of sodium bicarbonate and sodium chloride reabsorption, as well as the final stage of glucose reabsorption. It is also the exclusive site for transporting amino acids and significant anions, such as phosphate and citrate. The kidney is an organ that is actively involved in various metabolic processes such as the metabolism of vitamin D, gluconeogenesis, and clearance of drugs. It plays a critical role in maintaining the balance of potassium, sodium, and divalent cations through the distal convoluted tubule. This tubule reabsorbs approximately 10% of the filtered chloride and sodium, contributes to the secretion of potassium cations, and helps maintain the overall balance of magnesium and calcium in the body [58].

The inaugural model of toxicity was a research endeavor that involved a microdevice that has been designed to emulate the kidney-on-a-chip [59]. The researchers successfully produced the first multilayered microfluidic system. They used medullary collecting duct cells from mouse kidneys to replicate the mechanisms of renal filtration. The utilization of a microfluidic device facilitated the development of a biomimetic environment, resulting in an improved polarity of the inner medullary collecting duct. This enhancement was achieved through cytoskeletal reorganization and the stimulation of molecular transport via hormonal influences [18]. A microfluidic device was used to cultivate primary kidney proximal tubular epithelial cells under perfusion conditions in the experimental setting. Critical activities such as alkaline phosphatase activity, albumin transport, and glucose reabsorption could be studied using this microfluidic kidney-on-a-chip model, which replicated the fluidic flow seen in the human kidney's proximal tubule. The incorporation of a renal model, ideally in a three-dimensional (3D) configuration, into a platform that duplicates the microenvironment and composition of the kidney in its inherent condition constitutes a kidney-on-a-chip model. The kidney-on-a-chip model aims to facilitate the co-cultivation of diverse types of renal cells while upholding their functional interconnections, such as those between podocytes and glomerular vascular endothelial cells. Moreover, the kidney-on-a-chip model must exhibit the vital traits of endocrine functionality, metabolic activity, and functional transporter expression. The kidney-on-a-chip model serves a variety of purposes, including high-throughput screening for drug toxicity testing. It can be utilized to investigate the processes involved in glomerular filtration, assess pharmacokinetics, and improve the accuracy of drug dosage evaluations. By closely mimicking the human kidney, this model enables more accurate predictions of drug-induced nephrotoxicity, thereby decreasing the dependence on animal testing during preclinical trials. While models of the distal tubules and the proximal and glomerulus have been established, the ultimate objective of creating a comprehensive nephron-on-a-chip platform that integrates all of these elements has yet to be realized [59].

6.16 Brain-On-A-Chip

The brain, being a highly intricate organ comprising diverse cell types, exhibits significant genetic and functional disparities between humans and animals. Consequently, animal models only offer fundamental insights into brain functions and diseases [59]. The blood–brain barrier (BBB) is a sophisticated system of vascular endothelial cells that functions to segregate the central nervous system (CNS) from the systemic blood circulation, with the exception of designated regions. Notably, the BBB exhibits unique biochemical and physical characteristics compared to other endothelial cells, serving to protect neurons from hazardous agents while also regulating the entry of drugs. Therefore, it is imperative to thoroughly investigate the structure and operation of the BBB both in vitro and in vivo [24, 56]. To more accurately emulate the electrical, physical, and chemical conditions of the human brain, researchers have developed microfluidic brain-on-a-chip models [59]. Notably, recent advancements have successfully reconstructed essential aspects of the BBB on microfluidic chips, employing fluidic shear stress and thin membranes. These models incorporate co-cultured endothelial cells and astrocytes, overcoming limitations observed in previous models. Promisingly, they exhibit the potential to simulate BBB function effectively [24].

Various microfluidic brain models have been devised to investigate distinct facets of neurobiology. For instance, platforms have been created to study high-resolution axonal transport, monitor axonal growth, and assess the impact of extracellular matrix components on axon growth. Researchers have also constructed multilayered scaffolds, facilitated neural cell culture, and developed 3D architectures to mimic brain microenvironments. These models have been employed in studying neurodegenerative diseases, conducting drug tests, and even crafting brain tumor-on-a-chip models for therapeutic evaluations [59]. In addition, a synthetic microvasculature model of the blood–brain barrier (BBB) called SyM-BBB has been created using microfluidics technology. The model employs a plastic microfluidic chip with two compartments, one of which is designed to mimic the size of microcirculation. This unique design allows for separate manipulation of the apical and basolateral sides. Endothelial cells are placed on the apical side, while neuronal cells are placed on the basolateral side. Experimental assays using this chip have demonstrated its effectiveness in facilitating studies on drug delivery. Brain slices, particularly when cultured on microfluidic chips, function as valuable tools for conducting experiments on neuronal activity. Recent research has shown that brain-on-a-chip systems can overcome the difficulties of maintaining the viability and functionality of brain slices. These systems offer numerous advantages, such as enhanced oxygen and nutrient delivery to the slices, precise control over the flow of solutions and targeted drug administration and seamless integration with advanced optical and electrophysiological techniques. By combining the high-throughput capabilities of microfluidic technology with the tissue-like properties of in vivo brain slices, a perfect platform can be created for toxicology studies and conducting large-scale drug screening [24].

6.17 Heart-On-A-Chip

Cardiovascular diseases (CVDs) rank as the leading cause of global mortality. However, the use of animal models for studying cardiac toxicity and drug efficacy poses limitations due to their differing heart rates and QT interval variations from humans. To address these challenges, researchers have turned to developing human cardiac Organ-on-a-Chip (OOC) models. Creating heart-on-a-chip platforms presents significant hurdles, particularly in mimicking the dynamic microenvironments of cardiac tissue and studying both its contractility and electrophysiological responses concurrently. Nevertheless, some studies have made progress in cardiac tissue engineering and heart-on-a-chip platforms, enabling the replication of cardiac tissue microenvironments and the investigation of contractility and electrophysiological responses [15].

The myocardium, a crucial component of the heart, plays a vital role in assessing drug effects and heart pumping through the beating of cardiomyocytes (CMs) [18]. For these highly polarized, contracting cells to continue operating normally in terms of contractility and electrophysiology, they need an external physical–chemical environment. To successfully copy heart tissue on a tiny chip, many experimental factors must be brought together. These include flow rate, microscale topography, elastic properties, cell adhesion, and electrical and biochemical regimes. Recent microfluidic chip experiments have successfully replicated the structure of cardiac tissue on a small scale to obtain more pertinent morphometric, electrophysiological, and contractility data [56]. A planar electrode array and a PDMS microfluidic network were used to measure the extracellular potential of individual adult cardiomyocytes in the early stages of developing heart-on-a-chip technology. Other approaches included a microfluidic device with an electrode array to electrically measure metabolic profiles and optically assess cell contractility. Subsequent advancements introduced tissue-level heart-on-a-chip models, such as the fabrication of muscular thin films (MTFs) seeded with neonatal rat ventricular cardiomyocytes. These MTFs responded to electrical stimulation, exhibiting deflection that could be measured. When electrical stimulation was applied from the field electrode, the MTFs exhibited a curling motion. This enabled the measurement of cell deflection in response to the electrical stimuli [15]. A heart-on-chip methodology was reported by Nguyen and colleagues, enabling precise and controllable physiological mechanical stimulation. This approach holds great potential for generating functional cardiac patches in vitro using immature cardiomyocytes, offering a possible solution for replacing injured cardiac tissues. Additionally, researchers have developed a microfluidic device capable of generating cardiac-like flow within a closed culture system. This device offers advantages in terms of miniaturization and a low circulatory volume (2–3 μL) [67].

6.18 Skin-On-A-Chip

The skin, which is the largest organ, performs several crucial functions. These include acting as the primary protective barrier against environmental stressors such as chemical, physical, and biological agents. The skin also prevents dehydration and regulates body temperature. In the fields of pharmaceuticals, toxicology, and cosmetics, it is necessary to use skin models that accurately reflect the physiology of human skin. These models help identify potential hazards and assess the effects of new compounds on the skin. The skin is a convenient target for testing various stress factors under different conditions. However, not all in vivo and in vitro models effectively represent human skin [59]. Therefore, there is a need to develop better alternatives that reduce the use of animals and improve the relevance to human skin. The design of a skin-on-a-chip must possess the ability to create an interface between air and liquid, allowing the uppermost layer of skin to be exposed to air while the dermal layer remains in contact with a medium or, ideally, a vascular system. Currently, skin-on-a-chip models employ primary cells, and induced pluripotent stem cells (iPSCs) to encompass all pertinent cell types or cell lines. Nevertheless, the complete cultivation of all skin cells from iPSCs, including dermal papilla cells responsible for hair growth, has not yet been accomplished. To fabricate biomimetic skin models, it is crucial to replicate the vasculature and blood circulation within a laboratory setting. By integrating tissue engineering techniques with microfluidic technology, it becomes possible to achieve media perfusion, thereby fostering the creation of more pertinent skin models. Moreover, this approach offers valuable insights for drug testing purposes (Fig. 6.6, [72]). Wagner and his colleagues [73] have introduced a microfluidic system that employs a peristaltic micropump. This innovative platform facilitates the co-culturing of skin biopsies and human artificial liver microtissues. They observed crosstalk between the two cultures during long-term exposure to fluid flow over a period of 14 days. This platform was used to investigate tissue sensitivity to troglitazone, a pharmaceutical substance. Enhancing the physiological relevance of in vitro skin models involves incorporating both epidermal and dermal layers, unlike models that solely consist of the epidermal layer. Abaci and colleagues [74] designed a skin-on-a-chip platform capable of recirculating media at optimal flow rates, eliminating the need for external tubing or pumps. This platform enables the determination of the physiological residence time of blood in the skin tissue, ensuring the presence of drug concentrations that are pertinent to drug toxicity testing. Its primary application lies in the field of drug toxicity testing. The fusion of biosensors and biochips has facilitated the in situ and real-time tracking of skin tissue reactions to test substances. Alexander et al. [75] have engineered an intricate skin-on-a-chip system encompassing a sensor that scrutinizes the transepithelial electrical resistance of the reconstituted human epidermis. A platform of this kind has made it possible to monitor metabolic parameters and changes in the skin tissue for a certain period. Kwak and his colleagues created a microfluidic platform called skin-on-a-chip, where the dermal and epidermal layers were grown together with human umbilical vascular endothelial cells. When exposed to doxorubicin and UV irradiation, the scientists observed various immune responses, including an increase in the secretion of cytokines and the movement of neutrophils into the dermal layer [76, 77].

In a study the researchers successfully produced a tri-layered polydimethylsiloxane microfluidic "skin-on-a-chip" device equipped with two permeable membranes. This model exhibits the ability to cultivate a variety of human skin cells, comprising the endothelium, dermis, and epidermis. The independent microfluidic conduits within this mechanism facilitate the circulation of different types of media

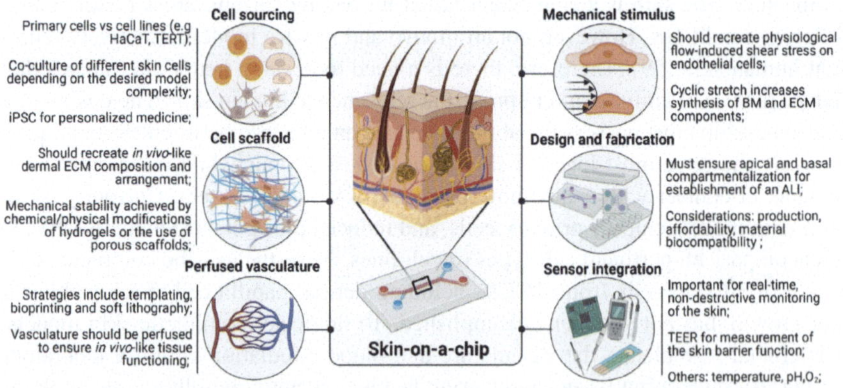

Fig. 6.6 The creation of platforms for biomimetic skin on a chip. A schematic illustration illustrating the key technical and biological aspects (cell source, cell scaffold, perfusion, cyclic stretching, design and manufacturing, and sensor integration) to take into account when creating physiologically realistic skin-on-a-chip (SoC) models. The many elements should be assessed with consideration for the particular application, equipment, and expertise that are accessible. BM stands for basement membrane; TEER stands for transepithelial electrical resistance; iPSC stands for induced pluripotent stem cells. ALI stands for air–liquid interface. Reprinted with permission from Ref. [72]. Copyright 2022, MDPI

at various flow velocities. By infusing tumor necrosis factor alpha into the channels, this structure effectively emulates cutaneous inflammation and edema. In addition, the authors have executed drug toxicity evaluations to investigate the effects of dexamethasone in diminishing inflammation and edema. A researcher has created a highly developed skin-on-a-chip model by utilizing a 3D co-culture method. This advanced model incorporates primary fibroblasts from the dermal layer embedded in a hydrogel matrix, resulting in a strong and authentic 3D dermal layer. Moreover, primary keratinocytes are cultivated on top of the collagen fibroblast layer to imitate the epidermal layer. The biochip comprises microfluidic channels that are coated with endothelial cells, facilitating the growth and distinction of the skin cells. Importantly, this model allows for the extended cultivation of skin cells for a duration of up to two weeks by perfusing through a microfluidic network. This remarkable development provides a powerful tool for studying skin biology and its various applications. A skin-on-a-chip platform has been introduced that replicated the effects of skin irritants on angiogenesis. More specifically, they activated the irritated keratinocytes through biochemical means to induce the production of vascular endothelial growth factors, thereby resulting in an increased sprouting of blood vessels. The researchers investigated the impact of steartrimonium chloride (a non-irritating compound) and sodium lauryl sulfate (a widely recognized irritant) on this platform. Recently, a simplified and efficient skin-on-a-chip model has been developed utilizing gelatin, which allows for the study of wound infection, pro-inflammatory responses of the skin, and drug screening. The model involved culturing keratinocytes on microchannels that were embedded in a gelatin matrix. Following six weeks of long-term culture, a multilayer structure that resembled the epidermal layer was formed. It is important to note that this model did not include fibroblasts or other cell types. Nonetheless, it functioned as a reliable model for studying pro-inflammatory responses to bacterial infection and conducting drug testing. In brief, a multitude of skin-on-a-chip prototypes have been established in the past few years. Nonetheless, the incorporation of a vascular network and immune cells is paramount for toxicity assessments and the examination of skin ailments in order to generate more dependable skin models that imitate the intricate 3D structure of human skin [59].

6.19 Vessel-On-A-Chip

The circulatory system in vertebrates comprises a vast network of blood vessels that interconnect all tissues and organs. Although the structure and size of these vessels may vary, the fundamental building blocks of the vascular system are endothelial cells (ECs), which function as a selective permeability barrier, separating tissues from blood. ECs play crucial roles in various physiological functions, including barrier function, vasodilation, vasoconstriction, angiogenesis, inflammation, blood clotting, and tumor metastasis. As a result, they have become a focal point in organ-on-a-chip research [67]. Blood vessels have a vital role in the transportation of blood throughout the human body. The arteries, which transport blood away from the heart, and the capillaries, which facilitate the exchange of water and chemicals between tissues and blood through the capillary endothelial barrier, are of utmost significance. On the other hand, veins are responsible for the transportation of blood back to the heart. It is essential to investigate the interaction between the vasculature and these carriers. Extensive research has examined the geometric characteristics of the vasculature, encompassing both simple and complex channels, as well as the accumulation of micro- and nanoparticles within these channels. Furthermore, it is crucial to consider the cyclic stretch and shear stress experienced by endothelial cells due to blood flow [15].

Precisely positioning endothelial cells on an existing supportive structure enables the development of Vascular organ-on-a-chip (OOC) models. These supportive structures can be created through methods such as sacrificial network formation, 3D printing, or injection molding for cell injection. Another approach involves embedding cells in a hydrogel, stimulating sprouting within larger channels. In the process of constructing human tissue-engineered blood vessels, mesenchymal stem cells (MSCs) are seeded in collagen. The vessel is then coated with endothelial cells and subjected to perfusion for a week. This process enables maturation and functionality, including vasoconstriction and vasodilation in response to drugs. Microvascular networks have also been achieved by incorporating endothelial cells and supporting fibroblasts into fibrin gels, and stimulating sprouting using chemical and mechanical stimuli. The ultimate goal of these approaches is the establishment of a perfusable microvasculature system. It should be noted that the application of flow regimens is vital for the successful creation of perfusable vascular networks [78]. A vessel-on-a-chip refers to a microfluidic apparatus specifically designed to emulate the properties of blood vessels. In numerous investigations, chambers constructed from PDMS and divided by a membrane are produced to cultivate vascular endothelial cells. These microfluidic chambers are interconnected with syringe pumps to regulate the flow rates, thereby enabling the examination of the shear stresses encountered by endothelial cells. Elastic membranes are employed to apply cyclic stretching strain to cells, wherein a groove layer and a vacuum pump contribute to the provision of suction pressure required for the stretching of the said membrane. Cells that are cultured atop the membrane undergo the process of cyclic stretching and subsequent relaxation. Scientists have successfully fabricated microfluidic chips that facilitate the establishment and examination of capillary networks employing human umbilical vein endothelial cells (HUVECs). These innovative devices showcase primary and bridge conduits brimming with fibrin hydrogels, facilitating the adhesion and translocation of HUVECs whilst inducing angiogenic procedures. Previous research has presented the inception of microvascular networks on chips by utilizing a fusion of fibrin gels, endothelial cells, and co-cultured fibroblasts. In general, the use of vessel-on-a-chip technology offers a highly effective method for examining the intricacies of the human vascular system in a controlled and consistent way. This not only allows for a greater understanding of the mechanisms behind diseases but also aids in the creation of innovative therapeutic interventions [15].

6.20 Cancer

Cancer, the second most common cause of death globally, continues to present considerable challenges in terms of understanding its biology and developing effective treatments [7, 8, 79–92]. The establishment of efficient tumor microenvironments in laboratory settings and the screening of anticancer drugs are critical to advancing cancer research. Despite the identification of several promising drug candidates in recent years, the efficacy of these drugs in clinical trials has proven to be disappointingly low, resulting in a failure rate of over 80% due to limited effectiveness or adverse reactions. Moreover, the cost of drug development has dramatically escalated, making the study of drug efficacy and toxicity a costly undertaking. The journey from identifying a potential drug compound to obtaining market approval is currently characterized by inefficiency. Despite the discovery of many hopeful drug candidates in recent years, the rate of success for these drugs in clinical trials continues to be disappointingly low, with more than 80% of candidates failing due to limited effectiveness or negative side effects. Moreover, the cost of developing drugs has significantly increased, and the study of drug effectiveness and toxicity remains a costly undertaking. The journey of transforming a potential drug from its initial identification to market approval is an inefficient process. Moreover, drugs that have already received approval can be withdrawn from the market due to unforeseen side effects, which pose significant health risks to patients. One major contributing factor to these challenges is the inability of animal experiments to accurately predict drug toxicity and side effects in humans, highlighting the limitations of current preclinical models used for disease and drug screening [93].

There are numerous compelling reasons to support the utilization of tumor-on-chip models as valuable instruments for conducting research on solid tumors. In essence, these reasons are associated with three pertinent attributes of solid tumors: (1) their three-dimensional composition, (2) their complexity and dynamic changes over time, and (3) their inherent inaccessibility and unobservability in animal models and human patients [94]. Traditional approaches to screening anticancer medications involve utilizing either in vitro cell cultures, such as 2D cell monolayers or 3D cell models, or in vivo animal models [93]. While animal models have contributed to our comprehension of intricate ailments like cancer and have provided valuable insights into drug impacts in living organisms, they are costly, time-consuming, and frequently fall short in accurately predicting human reactions during clinical trials. In practice, less than 8% of successful animal trials for cancer medications successfully transition into human clinical trials due to species-specific disparities in cell biology and physiology. Additionally, animal models offer limited possibilities for manipulating the examination of disease progression and therapeutic response. Although 2D monocultures are cost-effective, user-friendly, and high-throughput, their predictive capabilities are limited as they fail to replicate the natural physiology of human tissues. Transferring cells from 2 to 3D environments elucidates differences in gene/protein metabolism, response to stimuli, viability, proliferation, differentiation, morphology, cell behavior, expression, and drug sensitivity. The 3D structure of human organs is essential in developing and maintaining their specific functions, and the substrate topography, stiffness, and architecture in 2D cell cultures are vastly distinct from their in vivo counterparts. Furthermore, 2D cultures have the potential to result in loss of heterogeneity, genetic drift, and substantial alterations to cell phenotypes. 2D assays are limited in their ability to capture crucial elements of disease progression and treatment failure, including immune suppression and metastasis. The artificial in vitro conditions of 2D cultures restrict the exploration and targeting of important cell behaviors. Moreover, the absence of the tumor microenvironment and interactions between the tumor and surrounding tissues in two-dimensional cultures obstruct the delivery of drugs, potentially leading to the incorrect identification of drugs with limited effectiveness in living organisms. To address these limitations, scientists have created three-dimensional (3D) models that mimic the cellular environment and physiological conditions of native tissues, enabling more precise drug testing [95]. Conventional 3D cancer models created in vitro, while a valuable substitute for animal models are static systems lacking the dynamic component required for appropriate evaluation of drug efficacy and specificity. It has been acknowledged that the replication of a dynamic environment is a critical element in assessing the effectiveness of anticancer drugs. To overcome these limitations, the field of organ-on-a-chip (OOC) technologies has emerged, with the aim of reproducing the microphysiology and three-dimensional microstructure of human organs [93].

Organ-on-a-chip (OOC) devices combine microfluidic and 3D cell culture technologies, offering enhanced reproducibility, cost reduction, and high-throughput testing capabilities. They also provide improved recapitulation of the native microenvironment and ethical advantages. In the field of cancer research, OOCs are specifically designed for testing drug efficacy, modeling cancer metastasis, and creating personalized cancer models. These devices accurately recreate the complex microenvironments in which tumors develop.

6.21 Cancer-On-A-Chip

In the quest to understand and combat cancer, a groundbreaking approach called "tumor-on-a-chip" has emerged at the intersection of cancer organotypic culture and microfluidic devices. This cutting-edge technology allows scientists to recreate the intricate tumor microenvironment, shedding light on the behavior of cancer cells in a remarkably lifelike setting. By seamlessly integrating different physiological modules, including vasculature, these innovative tumor-on-a-chip models open up new avenues for exploring the dynamic interactions between cancer and other organs (Fig. 6.7) [96, 97]. The endeavor to comprehend and combat cancer has given rise to a pioneering technique known as "tumor-on-a-chip", which has emerged at the confluence of cancer organotypic culture and microfluidic devices. This state-of-the-art methodology empowers researchers to recreate the intricate tumor microenvironment, thereby illuminating the behavior of cancer cells in an exceedingly lifelike milieu. By seamlessly integrating diverse physiological modules, including vasculature, these innovative tumor-on-a-chip models open up novel avenues for investigating the dynamic interactions between cancer and other organs. A tumor is not a uniform mass of cells but instead, a complex ecosystem consisting of varied cell types, all intricately influenced by biochemical and biophysical cues. The genetic diversity found within tumors presents a significant challenge to effective treatment. However, experts have come to recognize that the tumor microenvironment also plays a crucial role in cancer initiation, progression, and the development of drug resistance. This understanding has led to a growing interest in targeting the tumor microenvironment as an independent and promising therapeutic strategy, regardless of the unique genomic aberrations present in each tumor. In this complex biological system, cellular communication occurs through a variety of mechanisms. The signaling network is intricately influenced by soluble factors, mechanical forces, cell-extracellular matrix adhesion, and cell–cell interactions in the interstitial fluid. Mechanical forces, including shear stresses, the composition and stiffness of the extracellular matrix, fluid flow, and organization, significantly impact cancer cell behavior. The utilization of tumor-on-a-chip technology offers unparalleled insights into the intricate interplay between cancer cells and their microenvironment within this dynamic context. The integration of malignant cells with neighboring tissues gives rise to a unique pathological "structure" that undergoes dynamic transformations as cancer progresses and responds to treatment. Similar to any living tissue, tumors rely on a vascular network to acquire oxygen and nutrients while eliminating metabolic waste. When adequate vasculature is lacking, solid tumors develop central necrosis due to limited oxygen diffusion, resulting in hypoxia, increased acidity, and waste buildup. Interestingly, cells located in the core of the tumor adapt to this challenging environment by adjusting their metabolism and entering a dormant state to maintain balance. This dormant population of tumor cells poses a significant obstacle to conventional treatments that primarily target rapidly dividing cells, such as chemotherapy and radiation. The employment of the tumor-on-a-chip technology not only furnishes a remarkable tool for investigating and comprehending the intricate dynamics but also harbors immense potential for augmenting preclinical appraisal of drug effectiveness. By scrupulously reproducing the tumor microenvironment and allowing for the amalgamation of multiple physiological constituents, including vasculature, tumor-on-a-chip models proffer a transformative platform for scrutinizing the intricate interplay between cancer cells and their surroundings. This innovative methodology is poised to revolutionize cancer research and impel the development of more efficacious therapeutic strategies for this pernicious ailment [95].

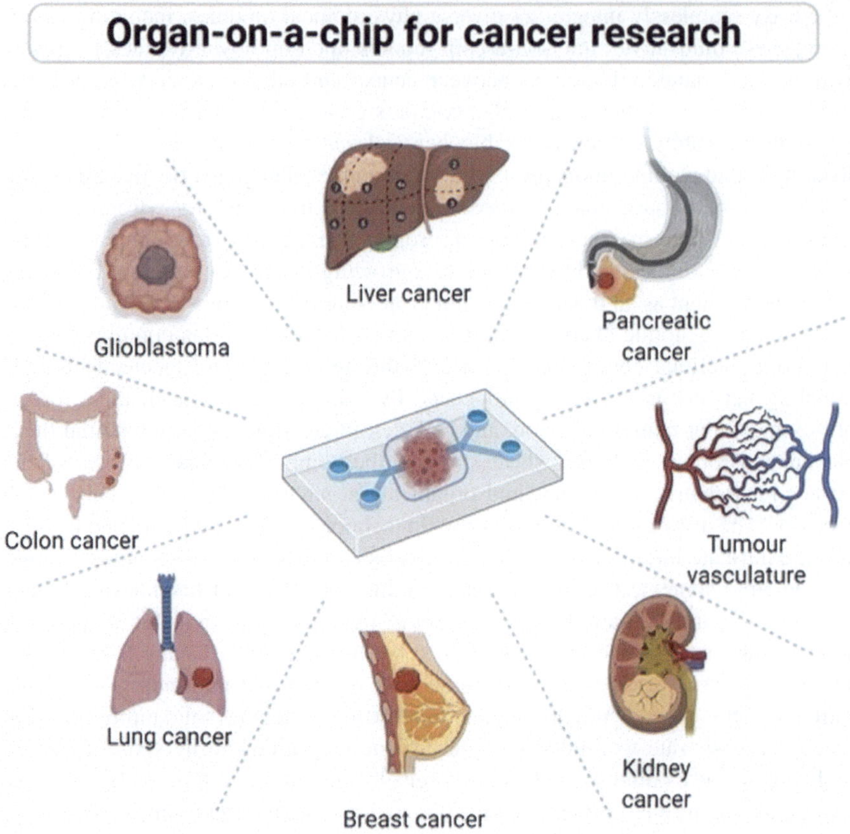

Fig. 6.7 Organ-on-a-chip for cancer research. Reprinted with permission from Ref. [97]. Copyright 2018, MDPI

6.22 Smart Materials for Cancer-On-A-Chip: Innovations in Manufacturing

The employment of microfluidic tumor-on-a-chip technology has been established as a highly valuable instrument in the area of drug research and toxicity testing. The construction of tumor-on-a-chip devices predominantly relies on Polydimethylsiloxane (PDMS) as the material of choice. PDMS offers a host of benefits, such as flexibility, biocompatibility, breathability, and optical transparency. These attributes enable real-time evaluation of cell behavior and response to treatments, facilitated by continuous microscopic examination of tissue constructs. PDMS-based devices, as opposed to substrates made of plastic or glass, provide a mechanical environment that more closely resembles soft tissues' physical properties, such as porosity and lower rigidity [93, 98]. Since PDMS can bind or absorb hydrophobic substances, its utility in chemical screening is restricted owing to its hydrophobic qualities. In order to overcome this challenge, Kang's research team devised a novel approach that involves the utilization of chemically bonded poly (methyl methacrylate) (PMMA) substrates and porous orbitally etched polyethylene terephthalate (PETE) membranes. This innovative microfluidic device effectively prevents the permeation of small lipophilic molecules. In drug testing experiments, the PMMA microfluidic device demonstrated more accurate results in assessing the cytotoxicity of vincristine when compared to the PDMS devices, as evidenced by the cultivation of human lung adenocarcinoma cells. This specific method shows great potential for creating tumor-on-a-chip devices by using porous track-etched membranes and a variety of thermoplastic materials. These devices allow for the development of 3D microstructures that accurately replicate physiological conditions at the organ level. Besides PMMA and PDMS, alternative materials can be used to construct microfluidic tumor-on-a-chip models. Gelatin, a biopolymer derived from collagen, is commonly employed in tissue engineering and drug transport models due to its ability to respond to cells and transport various biomolecules. Gelatin is well-known for promoting cell growth and can form polymers with polysaccharides, nucleotides, growth factors, proteins, and other polyionic complexes. This material exhibits great potential in drug transport models through targeted release of bioactive molecules and/or enabling controlled sustained release. Photocrosslinked gelatin methacrylate (GelMA) hydrogels have also been utilized in cartilage structure preparation, peripheral nerve regeneration, and corneal tissue engineering. Bacterial cellulose paper is a newly discovered material that presents several benefits, including the nanoscale properties of cellulose, its natural sourcing, cost-effectiveness, and biocompatibility. The nanofibers of bacterial cellulose can be easily processed and dried to form robust paper devices, which can be utilized for constructing models. A group of scholars have effectively created diverse microchannels using this paper-based equipment and established 3D paper vascularized breast tumor models, providing an uncomplicated and inexpensive option [93].

Cancer-on-a-chip models have broadened their scope of utility in examining the interrelationships between cancer and various organs. In their endeavor to comprehend the intrusion of cancer cells into the circulatory system (intravasation), Zervantonakis and his colleagues undertook a simulation of the three-dimensional interface between cancer cells and blood vessels. In their model, macrophages were integrated to investigate their role in promoting carcinoma intravasation. Through the use of live imaging, it was revealed that the damage to blood vessels caused by macrophages helped cancer cells enter into the bloodstream. The spread of cancer, known as metastasis, is an intricate process that relies on the interaction between cancer cells and the tissues they invade. This process involves the release of cytokines by the targeted organ in order to attract cancer cells to the site of metastasis. To investigate the "seed-soil" process, Jeon and colleagues devised a chip specifically designed for studying organ-targeted cancer extravasation in breast cancer. The chip encompasses units that imitate muscle and bone. By incorporating endothelial cells into a microvascular-like gel and embedding bone-mimicking stem cells, the researchers successfully lured extravasating cancer cells. This model replicates bone-specific extravasation of breast cancer cells, providing an avenue to explore the fundamental pathways implicated in cancer metastasis as well as screening potential anti-metastasis medications. Microfluidics has found numerous applications in modeling various processes involved in tumorigenesis, beyond intravasation and extravasation. These processes include adhesion, migration, invasiveness, and angiogenesis. Microfluidic chips have many advantages, including their capacity to link various modules for modeling multiorgan systems and their compatibility with online analytical modules that allow for real-time monitoring of organoid health. Zhang and colleagues successfully showed their "cancer-on-a-chip" platform, which effectively integrated several organoid models, such as cardiac organoids and liver cancer. The researchers have developed a complex multiorgan system that emulates the metabolic processes of doxorubicin (DOX) in living organisms. Within the microfluidic breadboard, they have integrated a variety of sensors, including physical sensors for monitoring pH levels, oxygen levels, and temperature, as well as biochemical sensors for detecting biomarkers. Additionally, an optical sensor is incorporated to image organoid morphology. Through the use of real-time sensing and facilitating interactions across many organs, they successfully monitored the toxicity and efficacy of DOX in a dynamic manner. This cutting-edge method allows for automatic, continuous monitoring of organoid reactions, removing the need for laborious sampling and reducing possible disturbances to the system. The microfluidic system successfully replicates both the anticancer effectiveness and the cardiac toxicity mediated by the liver metabolism of DOX. It should be noted that this brief overview of the "cancer-on-a-chip" concept does not provide exhaustive coverage of this extensively studied field. Interested readers are encouraged to refer to comprehensive reviews for more detailed information [96].

6.23 Diverse Tumor-On-A-Chip Models for Cancer Research

Tumor-on-a-chip models have become a potent instrument in the domain of cancer research, presenting an environment that is regulated to examine the demeanor and attributes of distinct cancer types. These novel frameworks have been employed to scrutinize a diverse array of cancers, imparting significant revelations regarding tumor biology, medication reactions, and probable therapeutic approaches. Various models of tumor-on-a-chip, which are created from human sources, have been developed to investigate the different characteristics associated with tumor progression, such as angiogenesis, drug response, growth, and metastasis. By utilizing advanced techniques in microfabrication, such as bioprinting, photolithography, and soft lithography, in conjunction with tissue engineering techniques and microfluidic, it becomes possible to replicate specific key elements of the tumor microenvironment within these chips. These elements encompass niche factors and biochemical gradients, dynamic interactions between cells and the extracellular matrix, as well as complex tissue structures comprised of both stromal and tumor cells. These innovative models have been employed to explore a broad spectrum of cancers, such as breast cancer, lung cancer, colorectal cancer, brain tumors, prostate cancer, ovarian cancer, pancreatic cancer, liver cancer, melanoma, leukemia, and osteosarcoma. Each cancer-on-a-chip model presents a distinct prospect to scrutinize the specific features, actions, and interplays of the corresponding cancer type, providing valuable insights into tumor progression, drug responses, and possible therapeutic approaches. By emulating the intricate tumor microenvironment and encompassing pertinent cell types and niche factors, these models enable researchers to gain an in-depth comprehension of the underlying mechanisms propelling each cancer type, resulting in advancements in personalized medicine and the creation of more effective treatment alternatives for patients [99–101].

6.24 Human-On-A-Chip

Researchers are currently working towards the development of a sophisticated multi-channel 3D microfluidic cell culture system that is capable of compartmentalizing microenvironments. This system would allow for the cultivation of 3D cellular aggregates that resemble multiple organs found in the human body. While most organ-on-a-chip models currently focus on culturing a single type of cell, which limits their ability to recreate systemic drug effects, an integrated cell culture analog known as μCCA has emerged as a promising solution. This innovative approach involves the inclusion of adipocytes in a 2D fluidic network that emulates blood circulation, facilitating nutrient delivery and waste removal, and drug-metabolizing liver cells, and lung cells. The advent of μCCA has paved the way for a realistic in vitro pharmacokinetic model that provides a biomimetic system capable of culturing multiple cell types to mimic in vivo conditions closely (Fig. 6.8 [58]). Zhang et al. [102] have further advanced this progress by implementing a microfluidic human-on-a-chip model that integrates four cell types to replicate key human organs, including kidney, lung, liver, and adipose tissue. Their emphasis on creating a standardized serum-free culture medium suitable for all cell types within the device is pivotal, as it ensures the maintenance of functional levels and enhances the sensitivity of the in vitro cultured cells. This accomplishment contributes to the validity of the human-on-a-chip system, enabling it to replicate physiological and metabolic reactions that are similar to whole organs when exposed to drugs in the microchannels [103–105].

6.25 A Rare Cancer and Human on a Chip

Rare cancers, characterized by an incidence rate of less than 150 new cases per million per year, pose unique challenges in comprehending their origins, progression, and management. Numerous body parts and organ systems, including the neurological, urogenital, respiratory, reproductive, endocrine, head and neck, and digestive

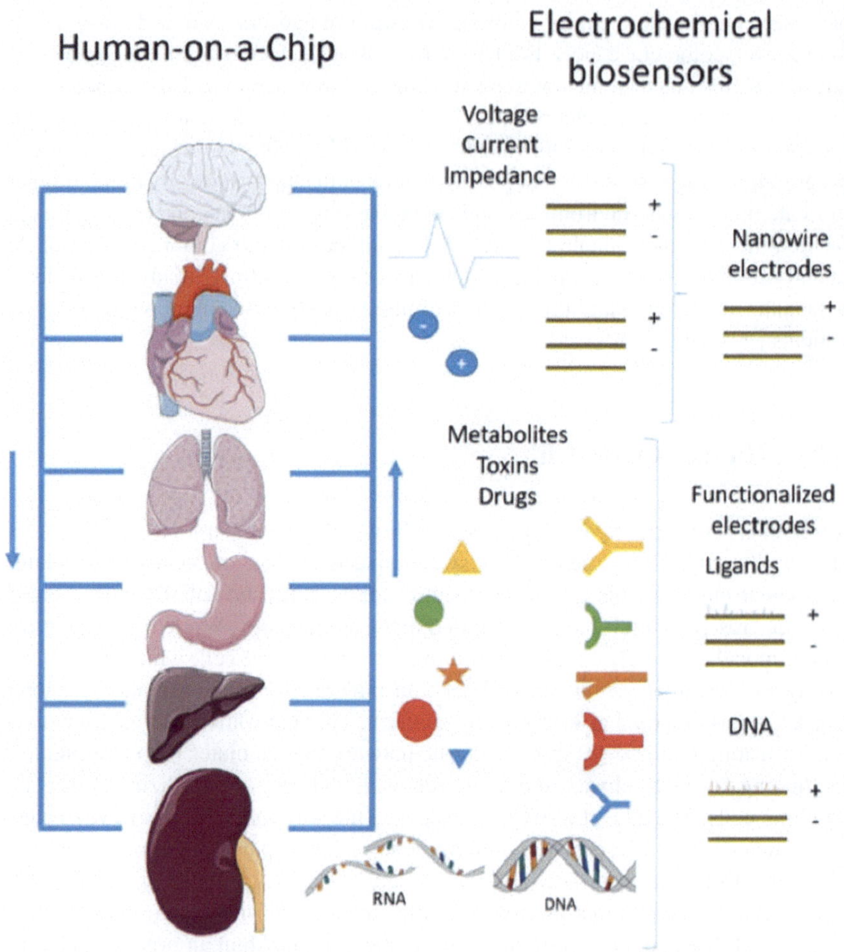

Fig. 6.8 Organs-sensors-on-a-chip microfluidic representation. Reprinted with permission from Ref. [58]. Copyright 2018, MDPI

systems, are susceptible to these cancers. Apart from rare histological variations, rare cancers also include molecular subtypes of common cancers, which significantly impact their behavior, response to treatment, and prognosis. For instance, pleomorphic lobular carcinoma, an aggressive form of lobular carcinoma, falls under the rare cancer category and exhibits a dismal prognosis. Rare cancers, although having low incidence rates, pose a formidable challenge for researchers to study, hindering the identification of effective therapeutic strategies within a reasonable timeframe. However, a beacon of hope shines through the emergence of Human-on-a-chip (HoaC) systems, which offer a promising avenue for investigating these cancers due to their modular and scalable nature. Thanks to their ability to accurately replicate essential aspects of tumor biology, such as the tumor microenvironment,

HoaC models show great potential. The tumor microenvironment, consisting of the signaling molecules, blood vessels, extracellular matrix, and inflammatory cells, plays a crucial role in cancer malignancy and the potential for metastasis. Microfluidic HoaC models can recreate specific components of the tumor microenvironment through innovative techniques such as the fabrication of microfluidic channels, micropatterning of the extracellular matrix, and the printing of cells and scaffolds in two or three dimensions [106].

6.26 Conclusion

In conclusion, organ-on-a-chip technology has emerged as a groundbreaking approach in the field of biomedical engineering. The fabrication of organs on a chip has opened up new avenues for studying complex physiological processes and disease mechanisms in a controlled and realistic.

Despite its numerous advantages, organ-on-a-chip technology is not without limitations. Challenges such as the complexity of recreating the entire human body and the need for further validation and standardization persist. However, the continuous advancements in design concepts, the integration of multiple organs, and the development of smart materials hold great promise for overcoming these limitations. The diverse range of organs-on-a-chip discussed in this chapter, including lung-on-a-chip, gut-on-a-chip, liver-on-a-chip, kidney-on-a-chip, brain-on-a-chip, heart-on-a-chip, skin-on-a-chip, and vessel-on-a-chip, highlights the versatility and potential of this technology in studying specific organ systems. Moreover, the application of organ-on-a-chip technology in cancer research, exemplified by cancer-on-a-chip and tumor-on-a-chip models, allows for a deeper understanding of cancer progression, drug responses, and personalized treatment approaches. Moving forward, the integration of multiple organ systems to create a human-on-a-chip platform holds immense promise for advancing our understanding of complex diseases and their interplay between different organs. The collaboration between researchers, engineers, and clinicians will be crucial in further refining and translating organ-on-a-chip technology into clinical applications. Organ-on-a-chip technology has revolutionized the way we study organs and diseases in the laboratory. With its potential to bridge the gap between traditional in vitro models and in vivo complexity, organ-on-a-chip has the power to accelerate drug discovery, improve personalized medicine, and ultimately lead to better healthcare outcomes for patients worldwide.

References

1. Sartipzadeh, O., et al.: Microfluidic-assisted synthesis and modeling of stimuli-responsive monodispersed chitosan microgels for drug delivery applications. Sci. Rep. **12**(1), 8382 (2022)

2. Sartipzadeh, O., et al.: Controllable size and form of droplets in microfluidic-assisted devices: effects of channel geometry and fluid velocity on droplet size. Mater. Sci. Eng. C **109**, 110606 (2020)
3. Rahmanian, M., et al.: A micropillar array-based microfluidic chip for label-free separation of circulating tumor cells: the best micropillar geometry? J. Adv. Res. **47**, 105–121 (2023)
4. Sartipzadeh, O., et al.: Microfluidic-assisted synthesis and modelling of monodispersed magnetic nanocomposites for biomedical applications. **9**(1), 1397–1407 (2020)
5. Ingber, D.E.J.D.: Developmentally inspired human organs on chips. **145**(16), dev156125 (2018)
6. Terrell, J.A., et al.: From cells-on-a-chip to organs-on-a-chip: scaffolding materials for 3D cell culture in microfluidics. J. Mater. Chem. B **8**(31), 6667–6685 (2020)
7. Naghib, S.M., Mohammad-Jafari, K.: Microfluidics-mediated liposomal nanoparticles for cancer therapy: recent developments on advanced devices and technologies. Curr. Top. Med. Chem. (2024)
8. Mohammad-Jafari, K., Naghib, S.M.: 3D printing of microfluidic-assisted liposomes production for drug delivery and nanobiomedicine: a review. Curr. Med. Chem. (2024)
9. Bhatia, S.N., Ingber, D.E.J.N.B.: Microfluidic organs-on-chips. **32**(8), 760–772 (2014)
10. Low, L.A., et al.: Organs-on-chips: into the next decade. Nat. Rev. Drug Discov. **20**(5), 345–361 (2021)
11. Shanehband, N., Naghib, S.M.: Microfluidics-assisted tumor cell separation approaches for clinical applications: an overview on emerging devices. Comb. Chem. High Throughput Screen (2024)
12. Shanehband, N., Naghib, S.M.: Recent advances in nano/microfluidics-based cell isolation techniques for cancer diagnosis and treatments. Biochimie **220**, 122–143 (2024)
13. Ronaldson-Bouchard, K., Vunjak-Novakovic, G.J.C.S.C.: Organs-on-a-chip: a fast track for engineered human tissues in drug development. **22**(3), 310–324 (2018)
14. Garshasbi, H.R., Naghib, S.M.: A review on nano/microfluidic devices for cell isolation techniques: recent progress and advances. Curr. Nanosci. **20**(6), 717–739 (2024)
15. Wang, Z., et al.: Organ-On-A-Chip Platforms for Drug Screening and Tissue Engineering, pp. 209–233 (2016)
16. Fuchs, S., et al.: In-line analysis of organ-on-chip systems with sensors: integration, fabrication, challenges, and potential. ACS Biomater. Sci. Eng. **7**(7), 2926–2948 (2021)
17. Tajeddin, A., Mustafaoglu, N.: Design and fabrication of organ-on-chips: promises and challenges. Micromachines **12** (2021). https://doi.org/10.3390/mi12121443
18. Wu, Q., et al.: Organ-on-a-chip: Recent breakthroughs and future prospects. **19**(1), 1–19 (2020)
19. Askari, E., et al.: Ultrasonic-assisted synthesis and in vitro biological assessments of a novel herceptin-stabilized graphene using three dimensional cell spheroid. Ultrason. Sonochem. **58**, 104615 (2019)
20. Ghazali, H.S., et al.: A high-absorbance water-soluble photoinitiator nanoparticle for hydrogel 3D printing: synthesis, characterization and in vitro cytotoxicity study. Sci. Rep. **13**(1), 8577 (2023)
21. Askari, E., et al.: Local delivery of chemotherapeutic agent in tissue engineering based on gelatin/graphene hydrogel. J. Market. Res. **12**, 412–422 (2021)
22. Abtahi, N.A., et al.: A comparative study on biopharmaceutical function of curcumin and miR-34a by multistimuli-responsive nanoniosome carrier: in-vitro and in-vivo. Front. Mol. Biosci. **9**, 1043277 (2022)
23. Ghazali, H.S., et al.: Lithography-based 3D printed hydrogels: from bioresin designing to biomedical application. Colloid Interface Sci. Commun. **50**, 100667 (2022)
24. Jiang, B., et al.: Organs on microfluidic chips: a mini review. **57**(3), 356–364 (2014)
25. Zhao, Y., et al.: Multi-organs-on-chips: towards long-term biomedical investigations. **24**(4), 675 (2019)
26. Saydé, T., et al.: Biomaterials for three-dimensional cell culture: from applications in oncology to nanotechnology. **11**(2), 481 (2021)

27. Rahmani Dabbagh, S., et al.: 3D bioprinted organ-on-chips. Aggregate **4**(1), e197 (2023)
28. Ma, C., et al.: Organ-on-a-chip: a new paradigm for drug development. Trends Pharmacol. Sci. **42**(2), 119–133 (2021)
29. Yang, Q., Lian, Q., Xu, F.J.B.: Perspective: fabrication of integrated organ-on-a-chip via bioprinting. **11**(3), 031301 (2017)
30. Miri, A.K., et al.: Bioprinters for organs-on-chips. **11**(4), 042002 (2019)
31. Rothbauer, M., et al.: Recent advances in additive manufacturing and 3D bioprinting for organs-on-A-chip and microphysiological systems. **10**, 837087 (2022)
32. Datta, P., et al.: 3D bioprinting for reconstituting the cancer microenvironment. **4**(1), 1–13 (2020)
33. Imparato, G., Urciuolo, F., Netti, P.A.: Organ on chip technology to model cancer growth and metastasis. Bioengineering **9** (2022). https://doi.org/10.3390/bioengineering9010028
34. Liu, X., et al.: Recent advances of organ-on-a-chip in cancer modeling research. **12**(11), 1045 (2022)
35. Niculescu, A.-G., et al.: Fabrication and applications of microfluidic devices: a review. Int. J. Mol. Sci. **22** (2021). https://doi.org/10.3390/ijms22042011
36. Elvira, K.S., et al.: Materials and methods for droplet microfluidic device fabrication. Lab Chip **22**(5), 859–875 (2022)
37. Ahadian, S., et al.: Organ-on-a-chip platforms: a convergence of advanced materials, cells, and microscale technologies. **7**(2), 1700506 (2018)
38. Scott, S.M., Ali, Z.: Fabrication methods for microfluidic devices: an overview. Micromachines **12** (2021). https://doi.org/10.3390/mi12030319
39. Cecen, B., et al.: Selection of natural biomaterials for micro-tissue and organ-on-chip models. J. Biomed. Mater. Res., Part A **110**(5), 1147–1165 (2022)
40. Gonçalves, I.M., et al.: Recent trends of biomaterials and biosensors for organ-on-chip platforms. Bioprinting **26**, e00202 (2022)
41. Nahak, B.K., et al.: Advances in organ-on-a-chip materials and devices. ACS Appl. Bio Mater. **5**(8), 3576–3607 (2022)
42. Soleymani, S., Naghib, S.M.: 3D and 4D printing hydroxyapatite-based scaffolds for bone tissue engineering and regeneration. Heliyon **9**(9), e19363 (2023)
43. Naghib, S.M., Zarrineh, M., Mozafari, M.R.: 3D printing chitosan-based nanobiomaterials for biomedicine and drug delivery: recent advances on the promising bioactive agents and technologies. Curr. Org. Chem. **28**(7), 510–525 (2024)
44. Mikaeeli Kangarshahi, B., Naghib, S.M.: Multicomponent 3D-printed collagen-based scaffolds for cartilage regeneration: recent progress, developments, and emerging technologies. Curr. Org. Chem. **28**(8), 576–594 (2024)
45. Kangarshahi, B.M., et al.: Bioprinting of self-healing materials and nanostructures for biomedical applications: recent advances and progresses on fabrication and characterization techniques. Bioprinting **38**, e00335 (2024)
46. Chliara, M.A., Elezoglou, S., Zergioti, I.: Bioprinting on organ-on-chip: development and applications. Biosensors **12** (2022). https://doi.org/10.3390/bios12121135
47. Moheb Afzali, A., Kheradmand, M.A., Naghib, S.M.: Bioreactor design-assisted bioprinting of stimuli-responsive materials for tissue engineering and drug delivery applications. Bioprinting **37**, e00325 (2024)
48. Pati, F. et al.: Printing three-dimensional tissue analogues with decellularized extracellular matrix bioink. Nature Commun., **5**(1) 3935 (2014)
49. Park, J.Y., Jang, J., Kang, H.-W.J.M.E.: 3D bioprinting and its application to organ-on-a-chip. **200**, 1–11 (2018)
50. Ravanbakhsh, H., et al.: Emerging technologies in multi-material bioprinting. Adv. Mater. **33**(49), 2104730 (2021)
51. Tabatabaei Rezaei, N., et al.: Recent advances in organ-on-chips integrated with bioprinting technologies for drug screening. Adv. Healthcare Mater. **12**(20), 2203172 (2023)
52. Knowlton, S., Yenilmez, B., Tasoglu, S.J.T.I.B.: Towards single-step biofabrication of organs on a chip via 3D printing. **34**(9), 685–688 (2016)

53. Kim, S., Takayama, S.J.K.R., Practice, C.: Organ-on-a-chip and the kidney. **34**(3), 165–169 (2015)
54. Wei, S., et al.: Organs-on-chips and its applications. **44**(4), 533–541 (2016)
55. Polini, A., et al.: Organs-on-a-chip: a new tool for drug discovery. Expert Opin. Drug Discov. **9**(4), 335–352 (2014)
56. Polini, A., et al.: Organs-on-a-chip: a new tool for drug discovery. **9**(4), 335–352 (2014)
57. Huh, D.D.: A human breathing lung-on-a-chip. Ann. Am. Thorac. Soc. **12**(Suppl 1), S42–4 (2015)
58. Sosa-Hernández, J.E., et al.: Organs-on-a-chip module: a review from the development and applications perspective. **9**(10), 536 (2018)
59. Azizipour, N., et al.: Evolution of biochip technology: a review from lab-on-a-chip to organ-on-a-chip. **11**(6), 599 (2020)
60. Thomas, D.P., et al.: Microfluidic gut-on-a-chip: fundamentals and challenges. Biosensors **13** (2023). https://doi.org/10.3390/bios13010136
61. Thomas, D.P., et al.: Microfluidic gut-on-a-chip: fundamentals and challenges. **13**(1), 136 (2023)
62. Kim, H.J., Huh, D., Hamilton, G., Ingber, D.E.: Human gut-on-a-chip inhabited by microbial flora that experiences intestinal peristalsis-like motions and flow. Lab Chip **12**(12), 2165–2174 (2012)
63. Kim, H.J., Ingber, D.E.: Gut-on-a-chip microenvironment induces human intestinal cells to undergo villus differentiation. Integrat. Biol., 5(9), 1130–1140 (2013)
64. Sung, J.H., Shuler, M. L.: A micro cell culture analog (microCCA) with 3-D hydrogel culture of multiple cell lines to assess metabolism-dependent cytotoxicity of anti-cancer drugs. Lab Chip, **9**(10), 1385–1394 (2009)
65. Villenave, R., Wales, S.Q., Hamkins-Indik, T., Papafragkou, E., Weaver, J.C., Ferrante, T.C., Bahinski, A., Elkins, C.A., Kulka, M., Ingber, D.E.: Human gut-on-a-chip supports polarized infection of Coxsackie B1 virus in vitro. PloS One **12**(2), e0169412, (2017)
66. Deng, J., et al.: Engineered liver-on-a-chip platform to mimic liver functions and its biomedical applications: a review. **10**(10), 676 (2019)
67. An, F., et al.: Organ-on-a-chip: new platform for biological analysis. **10**, ACI. S28905 (2015)
68. Liu, J., et al.: Design and Fabrication Of A Liver-On-A-Chip Reconstructing Tissue-Tissue Interfaces. **12**, 959299 (2022)
69. Kane, B.J., Zinner, M.J., Yarmush, M.L., Toner, M.: Liver-specific functional studies in a microfluidic array of primary mammalian hepatocytes. Anal. Chem., **78**(13), 4291–4298 (2006)
70. Lee, H., et al.: Cell-printed 3D liver-on-a-chip possessing a liver microenvironment and biliary system. Biofabrication **11**(2), 025001 (2019)
71. Cruz, D., et al.: The future of extracorporeal support. 36(4), S243–S252 (2008)
72. Zoio, P., Oliva, A.J.P.: Skin-on-a-chip technology: microengineering physiologically relevant in vitro skin models. **14**(3), 682 (2022)
73. Wagner, I.: A dynamic multiorgan-chip for long-term cultivation and substance testing proven by 3D human liver and skin tissue co-culture. Lap Chip, **13**(18), 3538–3547 (2013)
74. Abaci, H.E., et al.: Tissue engineering of human hair follicles using a biomimetic developmental approach. Nature Commun., **9**(1), 5301 (2018)
75. Alexander, Jr., F.A., Eggert, S., Wiest, J.: Skin-on-a-chip: transepithelial electrical resistance and extracellular acidification measurements through an automated air-liquid interface. Genes, **9**(2), 114 (2018)
76. Risueño, I., et al.: Skin-on-a-chip models: general overview and future perspectives. APL Bioeng. **5**(3), 030901 (2021)
77. Sutterby, E., et al.: Microfluidic skin-on-a-chip models: toward biomimetic artificial skin. Small **16**(39), 2002515 (2020)
78. Mandrycky, C.J., et al.: Organ-on-a-chip systems for vascular biology. J. Mol. Cell. Cardiol. **159**, 1–13 (2021)

79. Gooneh-Farahani, S., Naimi-Jamal, M.R., Naghib, S.M.: Stimuli-responsive graphene-incorporated multifunctional chitosan for drug delivery applications: a review. Expert Opin. Drug Deliv. **16**(1), 79–99 (2019)
80. Garshasbi, H.R., Naghib, S.M.: Smart stimuli-responsive alginate nanogels for drug delivery systems and cancer therapy: a review. Curr. Pharm. Des. **29**(44), 3546–3562 (2023)
81. Matini, A., Naghib, S.M., Mozafari, M.R.: Quantum Dots in Cancer Theranostics: A Thorough Review of Recent Advancements in Bioimaging, Tracking, and Therapy across Various Cancer Types. Current pharmaceutical biotechnology (2024)
82. Matini, A., Naghib, S.M.: Microwave-assisted Natural Gums for Drug Delivery Systems: Recent Progresses and Advances over Emerging Biopolymers and Technologies. Current Medicinal Chemistry (2024)
83. Shamaeizadeh, A., et al.: Smart Nanobiomaterials for Gene Delivery in Localized Cancer Therapy: An Overview from Emerging Materials and Devices to Clinical Applications. Current cancer drug targets (2024)
84. Mikaeeli Kangarshahi, B., Naghib, S.M., Rabiee, N.: DNA/RNA-based electrochemical nanobiosensors for early detection of cancers. Crit. Rev. Clin. Lab. Sci. 1–23
85. Mazidi, Z., et al.: Smart stimuli-responsive implantable drug delivery systems for programmed and on-demand cancer treatment: an overview on the emerging materials. Chem. Eng. J. **433**, 134569 (2022)
86. Abtahi, N.A., et al.: Multifunctional stimuli-responsive niosomal nanoparticles for co-delivery and co-administration of gene and bioactive compound: in vitro and in vivo studies. Chem. Eng. J. **429**, 132090 (2022)
87. Shahidi, M., et al.: HB5 aptamer-tagged graphene oxide for co-delivery of doxorubicin and silibinin, and highly effective combination therapy in breast cancer. Cancer Nanotechnol. **14**(1), 59 (2023)
88. Abtahi, N.A., et al.: Multi-sensitive functionalized niosomal nanocarriers for controllable gene delivery in vitro and in vivo. Cancer Nanotechnol. **14**(1), 1–18 (2023)
89. Abtahi, N.A., et al.: Development of highly efficient niosomal systems for co-delivery of drugs and genes to treat breast cancer in vitro and in vivo. Cancer Nanotechnol. **13**(1), 1–18 (2022)
90. Yazdan, M., Naghib, S.M., Mozafari, M.R.: Liposomal Nano-Based Drug Delivery Systems for Breast Cancer Therapy: Recent Advances and Progresses. Anti-cancer agents in medicinal chemistry (2024)
91. Akhlaghi, M., et al.: Synthesis and characterization of smart stimuli-responsive herbal drug-encapsulated nanoniosome particles for efficient treatment of breast cancer. **11**(1), 1364–1385 (2022)
92. Abtahi, N.A., et al.: Smart stimuli-responsive biofunctionalized niosomal nanocarriers for programmed release of bioactive compounds into cancer cells in vitro and in vivo. **10**(1), 1895–1911 (2021)
93. Liu, X., et al.: Tumor-on-a-chip: From bioinspired design to biomedical application. **7**(1), 1–23 (2021)
94. Trujillo-de Santiago, G., et al.: The tumor-on-chip: Recent advances in the development of microfluidic systems to recapitulate the physiology of solid tumors. **12**(18), 2945 (2019)
95. Hachey, S.J., Hughes, C.C.J.L.O.A.C.: Applications of tumor chip technology. **18**(19), 2893–2912 (2018)
96. Sun, W., et al.: Organ-on-a-chip for cancer and immune organs modeling. **8**(4), 1801363 (2019)
97. Gonçalves, I.M., et al.: Organ-on-a-chip platforms for drug screening and delivery in tumor cells: a systematic review. **14**(4), 935 (2022)
98. Sontheimer-Phelps, A., Hassell, B.A., Ingber, D.E.J.N.R.C.: Modelling cancer in microfluidic human organs-on-chips. **19**(2), 65–81 (2019)
99. Song, K., et al.: Diversity models and applications of 3D breast tumor-on-a-chip. Micromachines **12** (2021). https://doi.org/10.3390/mi12070814

100. Bērziņa, S., et al.: Technological advances in tumor-on-chip technology: from bench to bedside. Cancers **13** (2021). https://doi.org/10.3390/cancers13164192
101. Hachey, S.J., Hughes, C.C.W.: Applications of tumor chip technology. Lab Chip **18**(19), 2893–2912 (2018)
102. Zhang, Chi, et al.: Towards a human-on-chip: culturing multiple cell types on a chip with compartmentalized microenvironments. Lab Chip **9**(22), 3185–3192 (2009)
103. Syama, S., Mohanan, P.V.: Microfluidic based human-on-a-chip: a revolutionary technology in scientific research. Trends Food Sci. Technol. **110**, 711–728 (2021)
104. Abaci, H.E., Shuler, M.L.: Human-on-a-chip design strategies and principles for physiologically based pharmacokinetics/pharmacodynamics modeling. Integr. Biol.. Biol. **7**(4), 383–391 (2015)
105. de Mello, C.P.P., et al.: A human-on-a-chip approach to tackling rare diseases. Drug Discovery Today **24**(11), 2139–2151 (2019)
106. de Mello, C.P.P., et al.: A human-on-a-chip approach to tackling rare diseases. 24(11), 2139–2151 (2019)

The manufacturer's authorised representative in the EU is Springer Nature Customer Service Centre GmbH, Europaplatz 3, 69115 Heidelberg, Germany. If you have any concerns regarding our products, please contact ProductSafety@springernature.com

Printed and bound by CPI Group (UK) Ltd, Croydon, CR0 4YY
30/09/2026
02225592-0007